TREES
and
OTHER WOODY PLANTS OF MAINE

Their Occurrence and Distribution

An Annotated Catalog of the Woody Spermatophytes

Fay Hyland

and

Ferdinand H. Steinmetz

THE THORNDIKE PRESS · THORNDIKE, MAINE
and
University of Maine at Orono Press

Library of Congress Cataloging in Publication Data

Hyland, Fay, 1900–
　Trees and other woody plants of Maine, their occurrence and distribution.
　　Reprint of the 1944 ed. printed at the University Press, Orono, Me. which was issued as 2d series, no. 59 of University of Maine studies and published under title: The woody plants of Maine, their occurrence and distribution.
　　Bibliography: p.
　　Includes index.
　　1. Woody plants – Maine. I. Steinmetz, Ferdinand Henry, 1886— joint author. II. Title. III. Series: Maine. University. University of Maine studies; 2d series, no. 59.
QK484.M2H85 1978　　582'.15'09741　　78-18706
ISBN 0-89621-019-7
ISBN 0-89621-018-9 pbk.

CONTENTS

	PAGE
The White Pine—A Symbol	v
Acknowledgments	vi
Introduction	vii
Explanation of Catalog	ix
Catalog	1
Bibliography	58
Index to the families, genera, and common names included in the Catalog	65

PHOTO W. S. EVANS

THE CLOUD-SWEPT PINE

THE WHITE PINE—A SYMBOL

From both a historical and an economic viewpoint, the White Pine (*Pinus Strobus* L.) is without doubt the most famous of all the woody plants of the State of Maine. From the time of the early colonists, it has played an important role in the welfare of her people. The early claim to choice specimens of pine trees (the King's Pines) by the British Crown is well known. Until after the American Revolution every White Pine tree over two feet in diameter, growing in any part of the State (except those trees growing in areas previously granted to private persons) was the property of the British Crown, reserved for masts in the Royal Navy. Many of the trees were marked with the broad arrow, and all were protected by imposing heavy fines on the trespasser. The importance of the White Pine is indicated by the legal provisions found in the early charters and laws of the colonies, as cited in Francis Newton Thorpe's *The Charter of Massachusetts Bay*—1691 (3:1885-1886. 1909).

Maine has long been known as the Pine Tree State, principally because the White Pine was the chief source of lumber in early times. Exports, chiefly in the form of ton timber, were made to England from the early settlements. By the end of the nineteenth century the most valuable stands of White Pine timber had been exploited, lumbermen having already turned their attention elsewhere to meet the ever increasing needs in this country because of Western expansion.

Chests, cabinets, and other articles manufactured from the aged and mellow wood of the so-called "Pumpkin Pine" have long been treasured by their owners and eagerly sought by collectors.

As early as 1820 the White Pine tree became symbolic of Maine when, at the first meeting of the State Legislature, it was adopted as one of the emblems in the shield of the State coat of arms. Later, in 1895, the *"pine cone and tassel"* was legally adopted by the State Legislature as the floral emblem of Maine.

Poets have immortalized this tree. Thus we read of the "murmuring Pines and the Hemlocks," from Longfellow's *Evangeline,* and the "cloud-swept Pine-'erect' " from Whittier's beautiful song dedicated to the "lumbermen." John Kendrick Bangs in his poem entitled *"The Pine"* regards it in high esteem among its forest associates. Certainly this tree merits a place in the heart of every nature lover, especially if he be a native of Maine, where the White Pine has been appropriately chosen as symbolic of her people.

ACKNOWLEDGMENTS

The authors are indebted to many persons who have assisted in the research or in the preparation of the manuscript. Thanks are gratefully extended to Professor M. L. Fernald for his stimulating interest in the work and his readiness to assist in questions pertaining to nomenclature and identification of certain miscellaneous specimens; Dr. L. H. Bailey for determinations on the genus *Rubus;* Mr. E. J. Palmer on *Crataegus* spp. and certain *Quercus* hybrids; Dr. W. H. Camp on *Vaccinium;* Dr. C. R. Ball on *Salix;* the late Dr. K. M. Wiegand on *Amelanchier;* Mr. C. A. Weatherby for his many helpful suggestions regarding nomenclature, bibliography, and identification; Dr. Lyman B. Smith, Miss Marjorie W. Stone, and Miss Ruth D. Sanderson of the Gray Herbarium for their many courtesies; the late Mr. Arthur H. Norton for his personal interest in the work and valuable records on distribution; Dr. E. C. Ogden for advice on certain points pertaining to nomenclature; and to Professor William F. Scamman for critically reviewing the manuscript. It is to be understood that the taxonomic specialists consulted regarding certain specific plant groups are not to be held responsible in any way for the disposition of the groups in this publication.

Appreciation is expressed to the following persons for permission to use the herbaria indicated: Professor M. L. Fernald, Gray Herbarium; Mr. F. W. Hunnewell, New England Botanical Club; Dr. E. D. Merrill, Arnold Arboretum; Dr. Edward Wigglesworth, Boston Society of Natural History; and the late Mr. Arthur H. Norton, Portland Society of Natural History.

Financial aid granted by the Coe Research Fund of the University of Maine is gratefully acknowledged.

INTRODUCTION

The need for an annotated catalog of the woody plants of Maine has been evidenced by frequent requests from a large and diverse group of people whose interests in the subject are avocational as well as vocational. Even though there are several manuals, local floras, and periodicals devoted to the nomenclature, description, occurrence, and distribution of New England plants, those persons who seek specific information on Maine woody plants find the general sources inadequate or difficult to obtain. Furthermore, the information on occurrence and distribution is often of a general nature covering large areas and not applying specifically to Maine. With the need for a treatment on the occurrence and distribution of the woody plants of the State in mind, it seemed desirable to compile and publish this catalog.

Regarding the desirability of a critical survey of the plants of the State an interesting comment by Fernald (*Rhodora* 41:479. 1939) on a statement made by William Oakes in 1828 might be appropriately inserted here. In August, 1828, William Oakes, writing to his friend, Dr. James W. Robbins, said: "The greater part of July I have spent 'down East,' even as far as Quoddy Head which lieth more eastward than Eastport. I have seen there however but few plants new to N[ew]. E[ngland]. & am convinced that no great accessions to the N. E. Flora, and of absolutely new plants hardly any, are to be expected from the State of Maine." To this statement Fernald comments: "That was before my day, but as a school-boy I began the discovery in Maine of plants Oakes had never dreamed of and we now know from that state, so completely dismissed by Oakes from the botanical map, no less than 115 native species which are found nowhere else in New England."

One contributing cause to the variety of plant life in Maine is geological. The soils derived from the Ordovician, Silurian, and Devonian calcareous sediments of northern Maine are strongly differentiated from those of southern Maine which are derived from the acid (or only slightly calcareous) metamorphics and acid igneous rocks of undetermined geologic age (*Preliminary Geologic Map of Maine,* 1933, by Arthur Keith, State of Maine Geologic Survey). Many species of plants are known from the St. John valley which are quite absent from the acid soils to the south and southeast (Fernald in *Rhodora* 20:90. 1918). Other factors accounting in part for the diverse flora of the State are geographical position (especially with respect to the proximity to the ocean), and the various features of climate. Because of the influence of these and other factors, plant zones may be recognized in the State. A map dividing the State into zones of similar growing conditions based upon temperature (frost-free season) is shown by Clapp (*University of Maine Studies,* Second Series, No. 28:21. 1933).

This publication is an annotated catalog of the woody plants found within the borders of the State. In addition to the native trees, shrubs, and woody vines, it includes those exotics which have escaped and persisted without cultivation. In other words, those woody plants have been included which because of their persistence have come to be regarded as a part of our flora. The scope of this work has been extended in many cases to include street trees.

In order to make this work comprehensive, it was necessary not only to review the literature but also to obtain additional data from numerous specimens housed in selected herbaria of New England. The principal herbaria consulted were those of the New England Botanical Club, Gray Herbarium, Arnold Arboretum, Boston Society of Natural History, Portland Society of Natural History, and University of Maine. In addition, several private collections were examined. Further, it was found necessary to supplement these data with actual field records. Accordingly, a field survey was organized, and the State was covered systematically, making complete lists of all entities in representative areas. The survey was begun in 1933 and completed in 1939 (working summer months only). During the course of the survey, some 18,000 records were collected and listed on filing cards. These records, along with those previously tabulated from the herbaria, were plotted on Maine maps. In addition, authentic reports obtained from botanical publications and local residents were plotted on the maps. By the use of appropriate symbols these records were kept separate from those which were based upon specimens. The above data form the chief basis of the Catalog.

Immediately preceding the Catalog proper is an explanation of the method of cataloging and the terminology used. Included are photographic reproductions of typical habitats (Figs. 1-8) and a map of the State of Maine (Fig. 9).

Following the Catalog is a bibliography consisting of an annotated list of references especially applicable to Maine flora. Those references marked with an asterisk (*) contain the earliest known reports of plants of the State. Concluding the publication is an index to the families, genera, and common names of woody plants of the State.

EXPLANATION OF CATALOG

The following explanatory notes are given as an aid in the use of the Catalog. The arrangement of families, with few exceptions, follows that of Robinson and Fernald in *Gray's New Manual of Botany,* ed. 7. Genera, species, varieties, and forms are arranged alphabetically, and hybrids (those not treated as species) follow the parent occurring first in the list. Hybrids treated as species are preceded by the symbol "×."

Nomenclature is in accordance with the latest rules embodied in the International Code. Because of the general use of Gray's Manual, this text has been adopted as a basic reference; and, for the most part, the Manual names are employed in the Catalog. Entities which are not included in the Manual are usually followed by references to the original description, new combination, author citation, or to other sources of information. Entities which appear in the Catalog under names different from those of the Manual have the Manual names listed in italics as synonyms. In certain cases the spelling of technical names differs from that of the Manual. For instance, recent writers (*Rhodora* 19:70. 1917; 42: 94-95. 1940; 43:220-222, 556. 1941) have pointed out that *pennsylvanicus* (*a, um*) should, in most cases, be spelled with a single *n* in the first syllable. *Fraxinus pennsylvanica* appears to be an exception.

Differentiation between woody and non-woody plants is difficult since there is no clear-cut line of separation between them. However, the woody members are usually regarded as plants with ligneous, perennial (biennial in *Rubus*) stems which increase in diameter each year by formation of annual rings. One might immediately recall many exceptions to this definition. The Lambs-quarters (*Chenopodium*) and Sunflower (*Helianthus*) are annuals which, under favorable conditions, become definitely woody. In fact, many herbaceous plants become quite woody at base. On the other hand, the Creeping Snowberry (*Chiogenes*), Twinflower (*Linnaea*), and Dwarf Mistletoe (*Arceuthobium*) barely come within the category of woody plants, but are included since they are usually expected in woody plant lists. Justification of the inclusion of these plants as woody and the exclusion of such plants as Speedwell (*Veronica officinalis*) and others (especially the many tufted perennials) would be difficult. In view of the difficulties involved, it was deemed logical to consider only those plants commonly found in manuals dealing exclusively with woody plants. Consequently Rehder's *Manual of Cultivated Trees and Shrubs* was in general chosen as a basis for determining whether or not a particular species is woody.

There appears to be no uniformly adopted plan for hyphenating and capitalizing common plant names which are composed of compound words. Thus the common name of *Hamamelis virginiana* is written: "Witchhazel" (H. L. Shantz et al, Forest Service Tree Name Committee, *Approved Changes in Sudworth's Check List,* U. S. Forest Service, 1940) ; "Witch-Hazel" (Rehder's *Manual of Cultivated Trees and Shrubs,* ed. 2, 1940) ; and "Witch-hazel" (Robinson and Fernald, *Gray's New Manual of Botany,* ed. 7, 1908). The practice generally followed in this Catalog is the decapitalization of the second part of such hyphenated names. Free use has been made of the hyphen in such cases to indicate that the second part of the name, when used in its proper sense, belongs to another plant. Thus Witch-hazel is not a true Hazel (*Corylus*) nor is it even closely related, botanically.

The order of counties follows a geographical arrangement from north to south and east to west, i.e., Aroostook, Penobscot, Piscataquis, Somerset, Franklin, Oxford, Washington, Hancock, Waldo, Knox, Lincoln, Kennebec, Androscoggin, Sagadahoc, Cumberland, and York. A map showing counties and minor civil divisions is included in the publication (immediately preceding the Catalog) for those not familiar with the State. Ranges are listed according to counties or fractional parts thereof or, when practical, by latitude and longitude. When not otherwise stated, the county records are based upon several or many stations. Reports are always indicated as such, but dubious records and unauthenticated reports have for the most part been omitted.

The mere fact that a particular plant is not listed in the Catalog as occurring in a given county (especially if it is common or abundant elsewhere) does not necessarily mean that it is absent there—but, rather, it indicates that specimens were not seen in connection with this survey. As the result of a study made of all available data compiled previous to this survey, it became evident that certain counties had not been uniformly represented either by collections or reports, indicating that botanical exploration had been more exhaustive in certain areas than in others. For instance, the flora of Mt. Katahdin has been well represented in herbaria for a long time, and duplicate sheets from this region are not uncommon; but no similar representative lists of the flora of certain other parts of the State are available. Another factor accounting for uneven representation of the flora of a given locality is the tendency to collect only the rare plants and to neglect those more abundant. Black Ash, for example, is known to be common and locally plentiful throughout the State; but from the number of reports and herbarium specimens available, it would appear to be rare. The common plants are often ignored by the collector; but when he does make a collection, it is frequently those plants which are especially large and showy or abnormal which attract his attention. Recently "mass collections" (proposed by Edgar Anderson, *Ann. Mo. Bot. Gard.* 28:287-292. 1941) are being made by some botanists in order to "bring into the herbarium information which now we can get only in the field."

The terminology used in the Catalog in connection with range, distribution, and habitat needs little explanation. The term "throughout" means that the plant has been found in every county of the State, usually in several to many places. When the plant is not found throughout, but is of local or restricted occurrence, the counties or fractional parts occupied are indicated. All statements on occurrence and distribution apply strictly to Maine unless otherwise indicated.

Following is a summary of the entities listed in the Catalog:

```
Families .........................................49
Genera ..........................................117
Species .........................................366
Varieties and named forms (exclusive of those which
    are the sole representatives of species in the State)..111
Hybrids .........................................36
Whole number of different plants (species, varieties
    and named forms, and hybrids) listed..................513
```

The illustrations on the following pages, xiii to xviii, appear with the kind permission of the Maine Forestry Department, Augusta, Maine.

───────────────

ATLANTIC WHITE-CEDAR
One-half natural size.

JACK PINE
Leaves, Cone and Seed.

BALSAM FIR
Branchlet and cone.

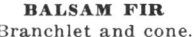

COMMON JUNIPER
Leaves and fruit.
One-half natural size.

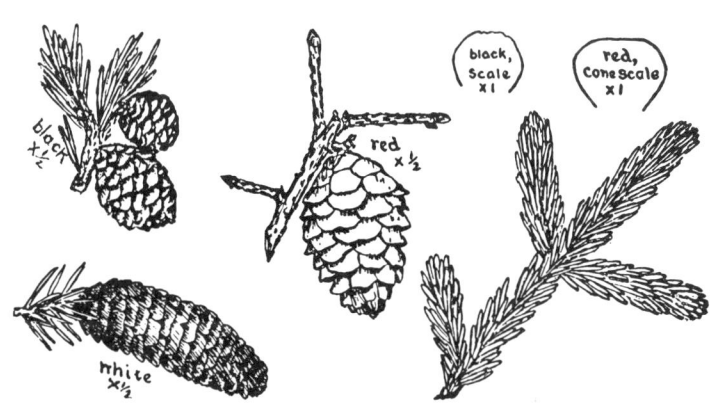

SPRUCES
Twigs, cones and cone-scales.

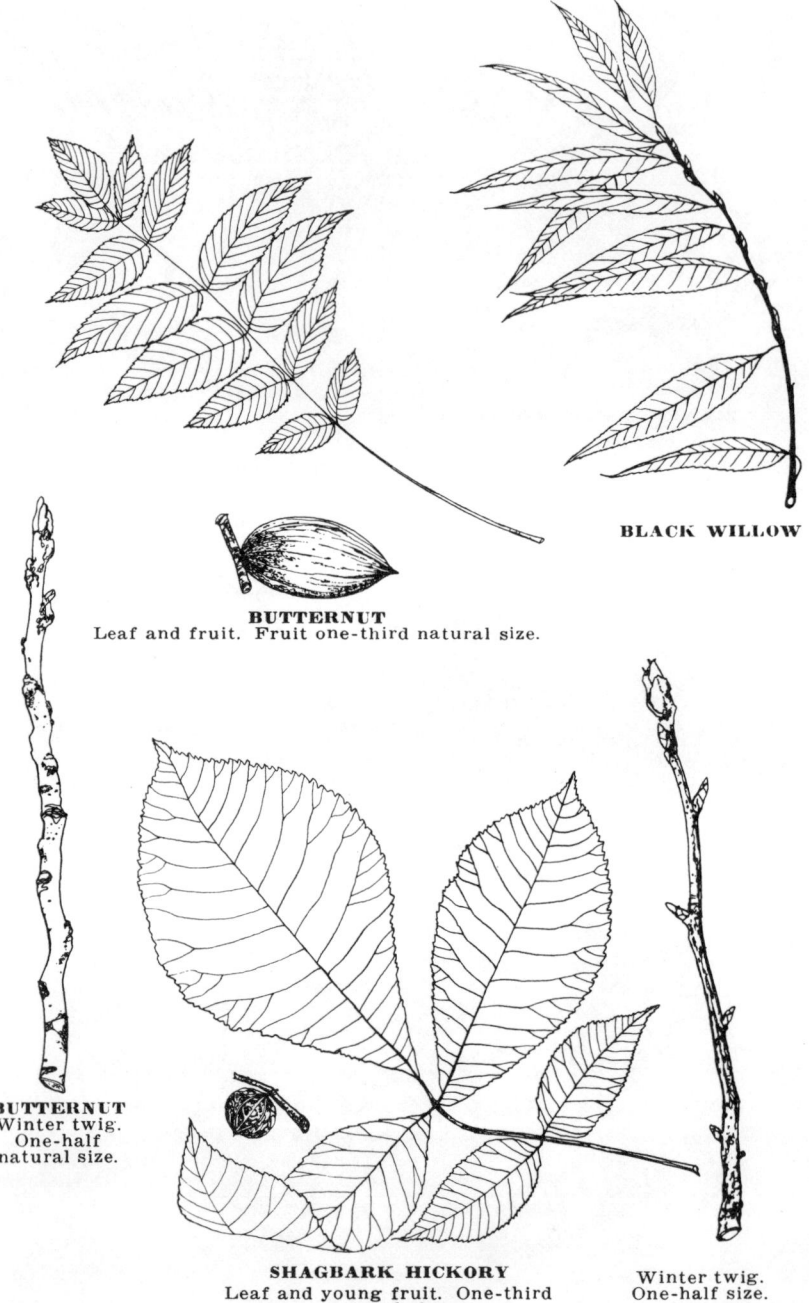

BLACK WILLOW

BUTTERNUT
Leaf and fruit. Fruit one-third natural size.

BUTTERNUT
Winter twig.
One-half
natural size.

SHAGBARK HICKORY
Leaf and young fruit. One-third
natural size.

Winter twig.
One-half size.

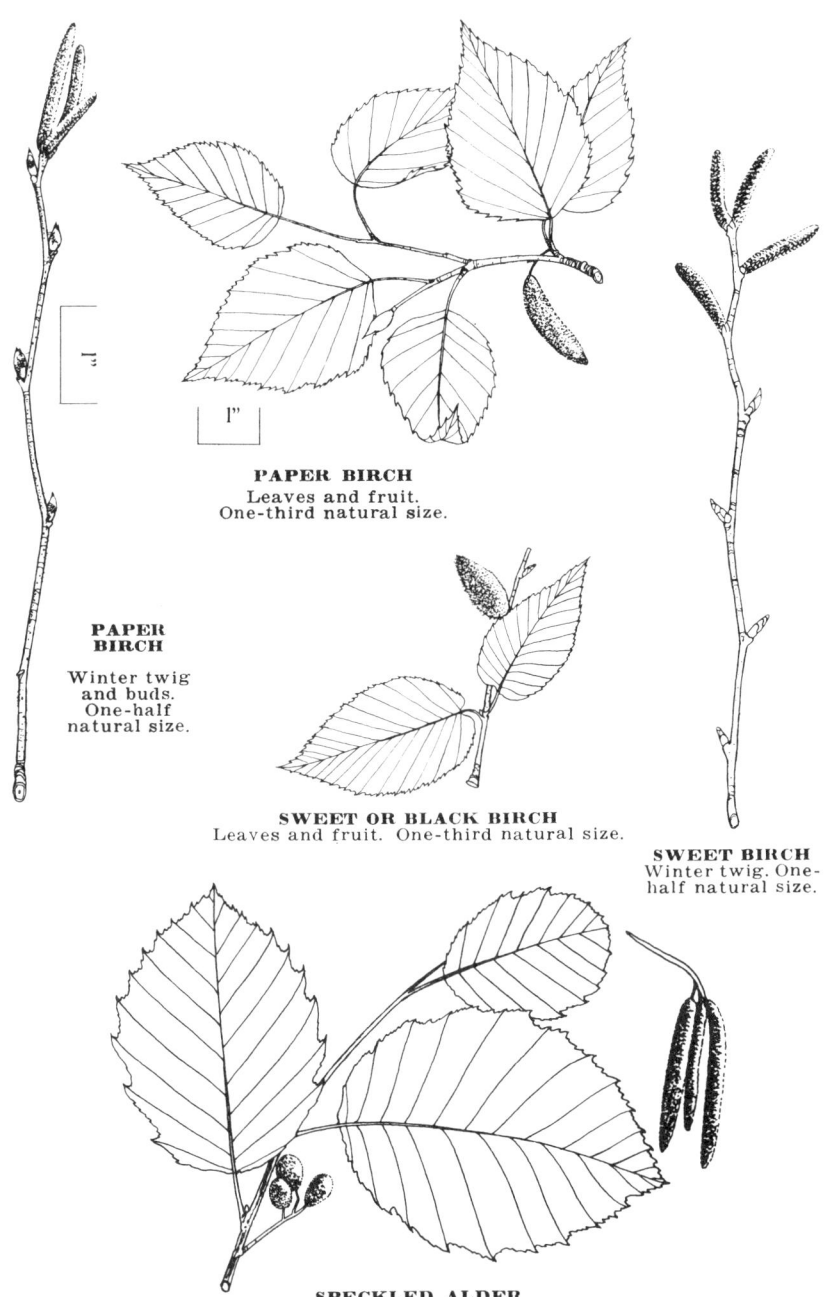

PAPER BIRCH
Leaves and fruit.
One-third natural size.

PAPER BIRCH

Winter twig
and buds.
One-half
natural size.

SWEET OR BLACK BIRCH
Leaves and fruit. One-third natural size.

SWEET BIRCH
Winter twig. One-half natural size.

SPECKLED ALDER

Leaves, flowers and fruit. One-third natural size.

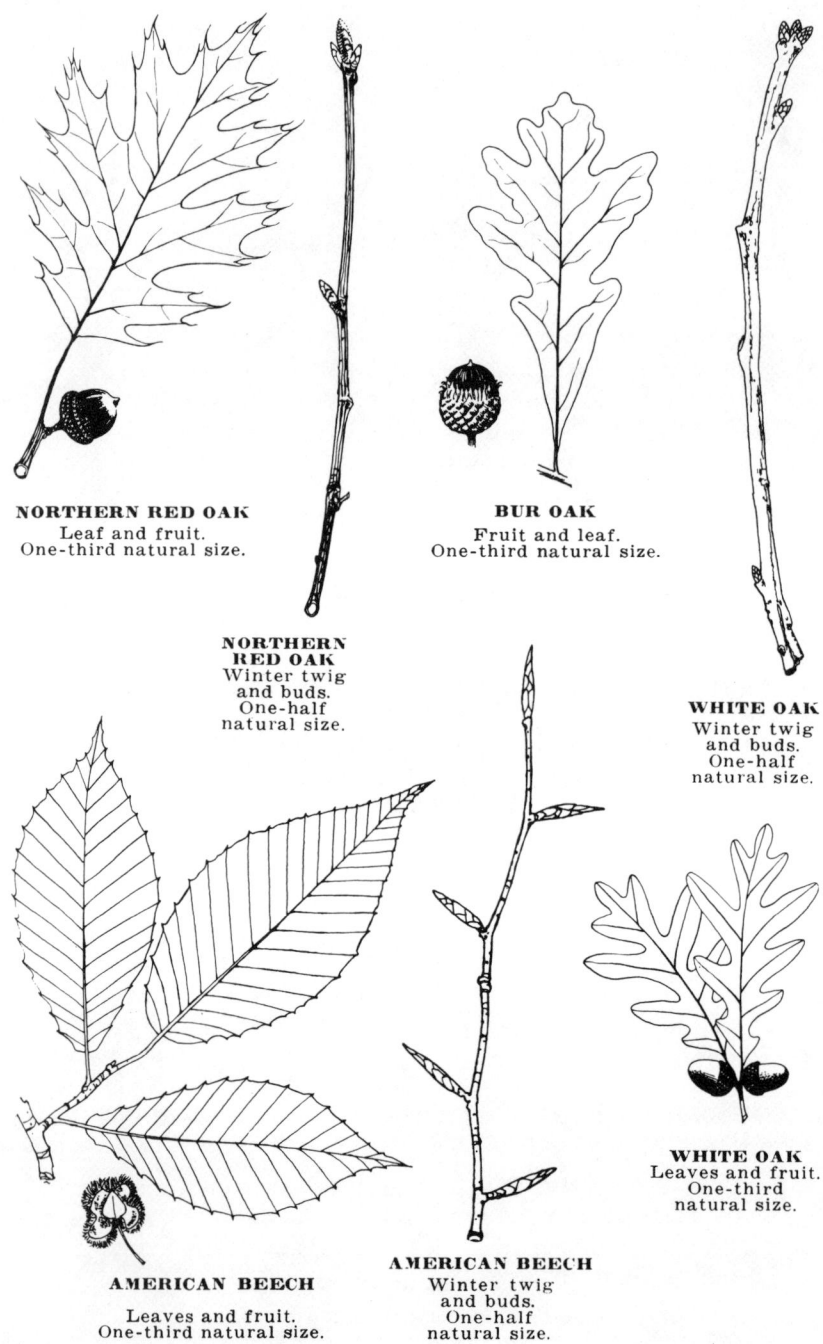

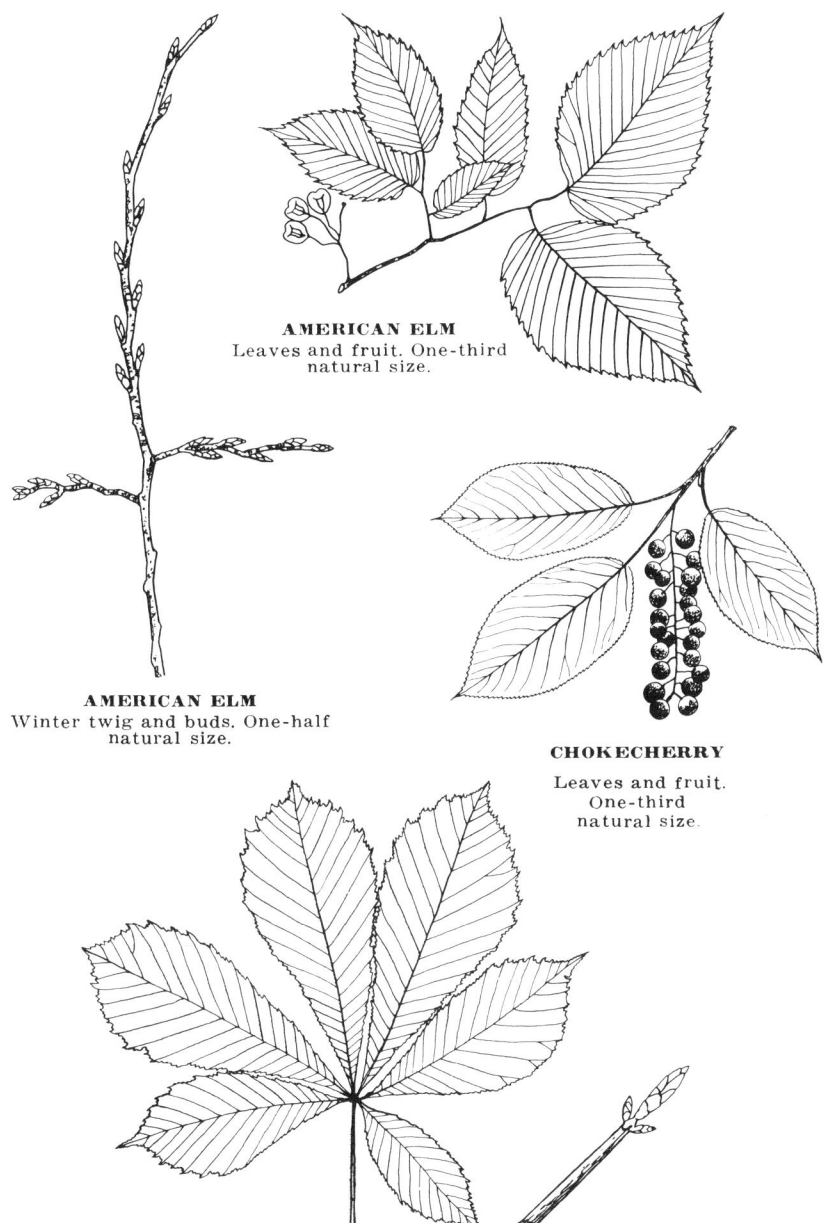

AMERICAN ELM
Leaves and fruit. One-third natural size.

AMERICAN ELM
Winter twig and buds. One-half natural size.

CHOKECHERRY
Leaves and fruit. One-third natural size.

HORSECHESTNUT
Leaf with leaflets. One-third natural size. Twig with buds. One-fourth natural size.

SUGAR MAPLE
Leaf and fruit.
One-third natural size.

RED MAPLE
Leaf and fruit. One-third
natural size.

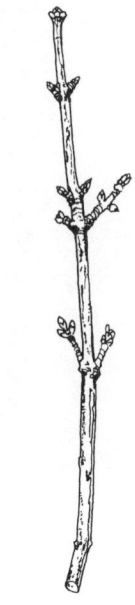

RED MAPLE
Winter twig
and buds.
One-half
natural size.

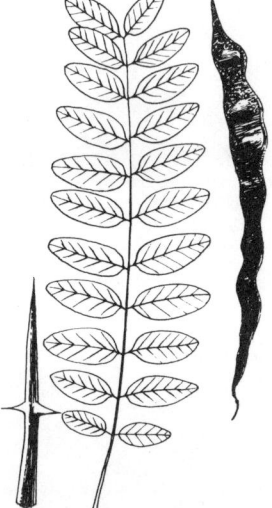

HONEYLOCUST
Spur, natural size, and leaf
with leaflets, and fruit.
One-third natural size.

BLACK ASH
Leaf with leaflets, and fruit. One-third natural size.

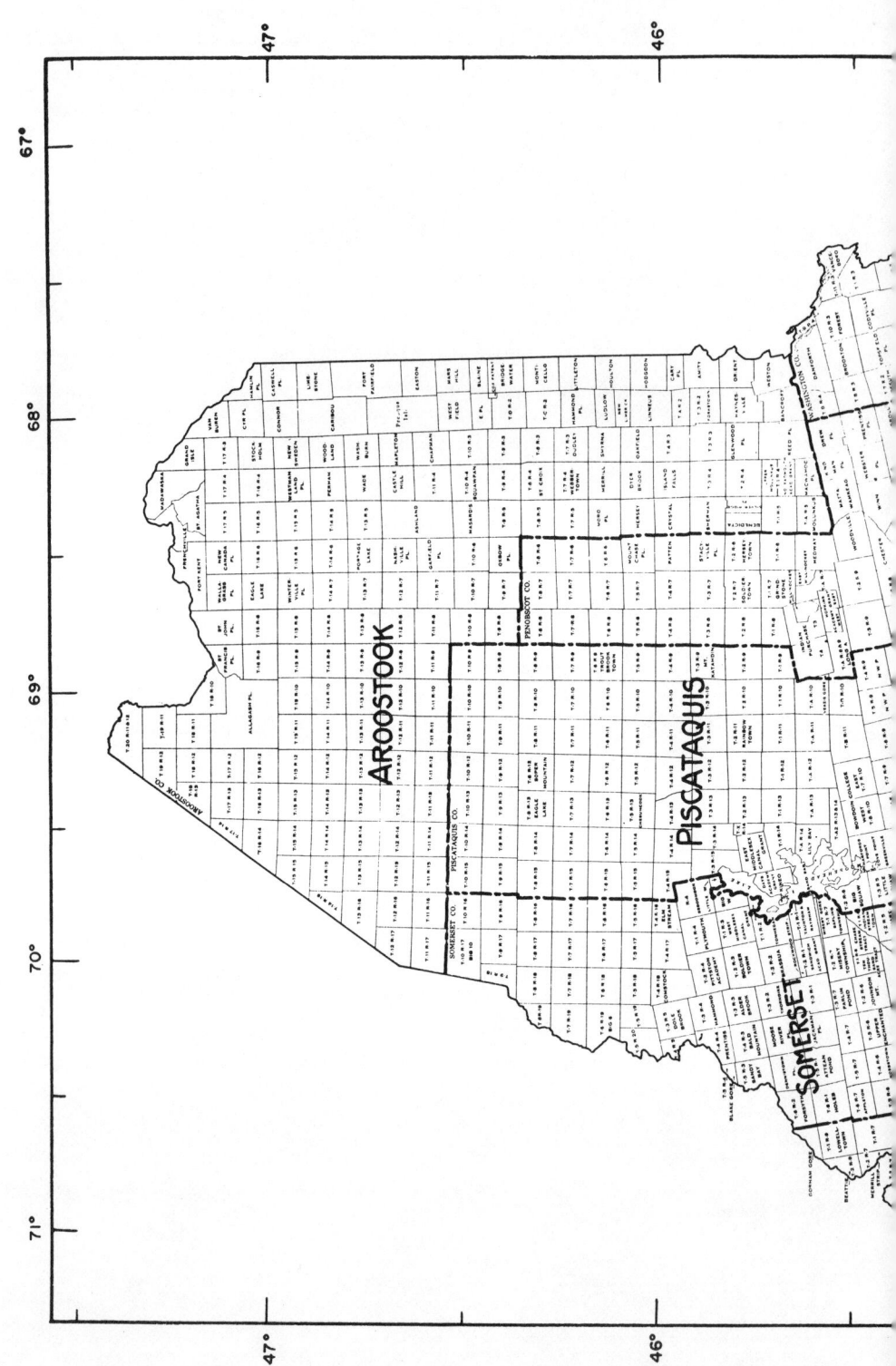

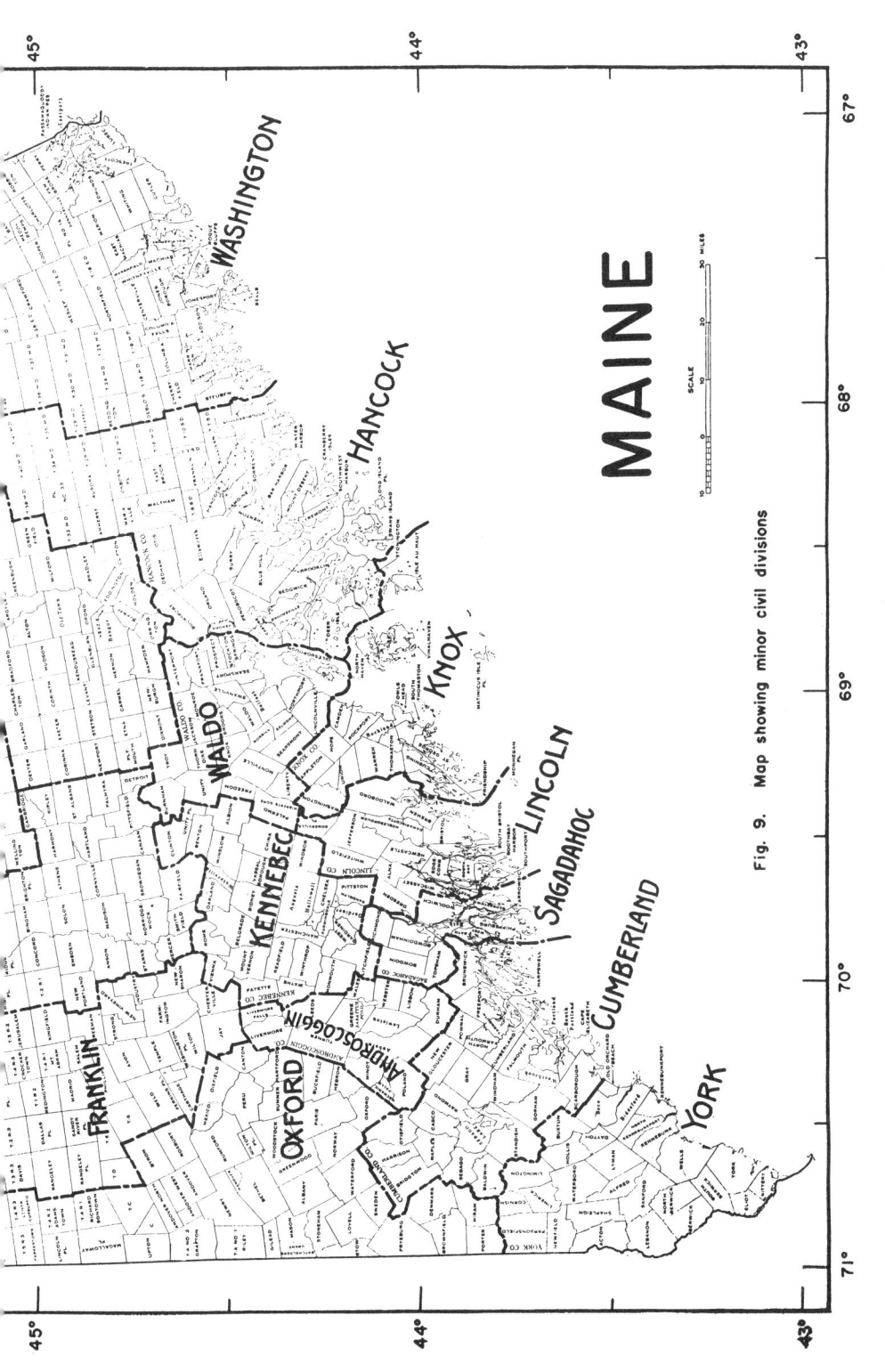

Fig. 9. Map showing minor civil divisions

CATALOG

GINKGOACEAE (Ginkgo Family)

Ginkgo. Ginkgo.

G. biloba L. Ginkgo, Maidenhair-tree.
See Rehd. Man. ed. 2:1. 1940.

Native to eastern China. Introduced as an ornamental and occasionally planted in southern Penobscot, eastern Oxford, southeastern Hancock, southwestern Waldo, northern Knox, central Kennebec, south central Androscoggin, eastern Cumberland, and eastern York Counties.

TAXACEAE (Yew Family)

Taxus. Yew.

T. canadensis Marsh. American Yew, Ground-hemlock.

Common and locally abundant. Damp, usually mucky soil in sparse or deep woods, mostly in the shade of conifers or yellow birch; occasionally on moist, rocky slopes at the higher elevations. Throughout.

PINACEAE (Pine Family)

Abies. Fir.

A. balsamea (L.) Mill. Balsam Fir.

Common forest tree, forming a considerable part of the forest on low swampy ground and in damp woods and mountain swamps; on well-drained hillsides associated with spruce; near the mountain summits reduced to a low, almost prostrate shrub forming dense mats. Throughout.

A. balsamea (L.) Mill. var. **phanerolepis** Fernald.
See Rhod. 11:201-203. 1909.

Infrequent. With the species. Oxford, Washington, Hancock, and Knox Counties. Perhaps more widespread than indicated by the records because of the inaccessibility of mature cones necessary for identification.

A. concolor (Gord.) Engelm. White Fir.
See Rehd. Man. ed. 2:17. 1940.

Indigenous to the Rocky Mountain region. Introduced and frequently planted in the urban centers in the portion of the State south of latitude 45.

Larix. Larch.

L. laricina (DuRoi) K. Koch. Tamarack, Eastern Larch.

Common and locally abundant. Chiefly in cold swamps but also on wet rocky hillsides and other seepy places. Throughout. Locally, but misappropriately, called "Juniper."

Picea. Spruce.

P. Abies (L.) Karst. Norway Spruce.

Native to Europe. Introduced and commonly planted as an ornamental and occasionally as a forest tree. Cultivated throughout, south of latitude 45, except southern Franklin, northern Oxford, Washington, Hancock, and Sagadahoc Counties.

P. glauca (Moench) Voss. White Spruce.
P. canadensis. See Rhod. 17:59-62. 1915.

Common and widespread, forming vast forests but less abundant in the hardwood region of Aroostook County. Shallow soils and rocks from sea level to the timber line in the mountains. Throughout, except Sagadahoc, and the western and central portions of Cumberland, and York Counties.

P. mariana (Mill.) BSP. Black Spruce.

Common and locally abundant. Cold bogs and mountain slopes; often on the hills and high banks of streams in Aroostook County. Throughout.

P. pungens Engelm. Blue Spruce.
See Rehd. Man. ed. 2:28-29. 1940.

Indigenous to the Rocky Mountain region. Frequently introduced and cultivated as an ornamental. Planted in northeastern Aroostook, southern Penobscot, southern Piscataquis, southern Somerset, southern Franklin, east central Oxford, southern Washington, southeastern Hancock, south central Kennebec, west central Androscoggin, eastern Cumberland, and southern York Counties.

P. rubens Sarg. Red Spruce.
P. rubra. See Rhod. 34:211. 1932.

Common and widespread, forming forests on shallow soils and in rocky upland woods from sea level to the timber line in the mountains. Throughout, but less abundant in the hardwood region of Aroostook County.

Pinus. Pine.

P. Banksiana Lamb. Jack Pine.

Infrequent and sporadic but locally plentiful. Barren, sandy, or rocky acid soil. Northwestern, central, and southern Penobscot, northern and central Piscataquis, central Somerset, northern Franklin, northwestern Oxford, southeastern Washington, east central and western Hancock, and north central (report) York Counties. Planted occasionally as an ornamental throughout its range. Specimens from Hancock and Washington Counties were reported as early as 1881 by C. G. Atkins of Bucksport, Maine (C. E. Bessey in the American Naturalist 15:316. 1881). (See article on soil type by Fernald in Rhodora 21:41-47. 1919.)

P. Mugo Turra. Mugo or Mountain Pine.
See Rehd. Man. ed. 2:41-42. 1940.

Native to Europe. Introduced and frequently planted. Southwestern Piscataquis, and throughout, south of latitude 45, except Washington, Lincoln, and Sagadahoc Counties.

P. nigra Arnold var. **austriaca** (Hoess) Aschers. & Graebn. Austrian Pine.
See Rehd. Man. ed. 2:42-43. 1940.

Native to Europe. Occasionally introduced as an ornamental. Planted in southeastern Aroostook, southern Penobscot, southern Somerset, western Hancock, east central Waldo, central Kennebec, and eastern Cumberland Counties, and probably elsewhere.

P. resinosa Ait. Red or Norway Pine.

Common and locally abundant, becoming less frequent northward. Dry woods and barrens on sandy or gravelly soil. Throughout, but apparently absent from northern Aroostook County.

P. rigida Mill. Pitch Pine.

Frequent and locally abundant, becoming sporadic northeastward. Barrens (in the southern part of the State) and sandy soil or rocks (northeastward along the coast). Throughout, south of latitude 44.5, except Waldo County.

P. Strobus L. Eastern or Northern White Pine, Pumpkin or Soft Pine.

Common and locally abundant. Mostly on fertile, well-drained soil of better quality than that occupied by other pines; on the banks of streams, river flats, or rarely in bogs (where it grows extremely slowly); also in rocky places at the higher elevations. Throughout, but more commercially important in the southern and eastern portions of the State.

P. sylvestris L. Scotch Pine.

Native to Europe. Introduced as an ornamental and widely cultivated. Planted throughout, but less frequently north and eastward.

Pseudotsuga. Douglas-fir.

P. taxifolia (Poir.) Britt. Douglas-fir.
See Rehd. Man. ed. 2:18-19. 1940.

Indigenous to western United States. Introduced as an ornamental and frequently planted. Southern Penobscot, southern Somerset, southeastern Oxford, west central Knox, eastern Cumberland, and eastern York Counties.

Tsuga. Hemlock.

T. canadensis (L.) Carr. Eastern Hemlock.

Common. Hilly or rocky woods and cool slopes; also on rather mucky soil near ponds and streams. Throughout.

TAXODIACEAE (REDWOOD FAMILY)

Sciadopitys. Umbrella-pine.

S. verticillata (Thunb.) Sieb. & Zucc. Umbrella-pine.
See Rehd. Man. ed. 2:48. 1940.

Native to central Japan. Occasionally introduced as an ornamental but not escaping. Planted in southeastern Hancock and York Counties.

CUPRESSACEAE (CYPRESS FAMILY)

Chamaecyparis. White-cedar.

C. thyoides (L.) BSP. Atlantic, Southern, or Coast White-cedar.

Locally rather plentiful. An area in York County approaching one thousand acres has been recently estimated (Rhodora 42:343-344. 1940). Quaking bogs or swampy ground bordering ponds or streams. Not known northeast of York County until collected by George B. Rossbach (Rhodora 38:453. 1936) from Waldo and Knox Counties in 1930 and 1931, respectively. Southeastern Waldo, northwestern Knox, and northwestern, central, and southeastern York Counties.

Juniperus. Juniper.

J. communis L. Common Juniper.

Infrequent and local. Dry, sandy, gravelly or rocky soil in open places. Southern Penobscot, central and southeastern Somerset, southern Franklin, east and west central Oxford, eastern and western Cumberland, and west central and south central York Counties.

J. communis L. var. **depressa** Pursh. Dwarf, Ground, or Prostrate Juniper.

Common and locally abundant, becoming infrequent northward. Poor sandy, gravelly and rocky soil in old pastures, along stone walls and other open places, characteristically forming large mats on the bare hillsides; also on exposed places in the mountains. T. 11, R. 8 and Hodgdon, Aroostook County, and throughout, south of latitude 46.

J. communis L. var. **saxatilis** Pall. Mountain Juniper.

J. communis var. *montana.* See Rehd. Man. ed. 2:62. 1940.

Local. Exposed rocky places along the coast and (rarely) on the mountains inland. East central Piscataquis, southeastern Washington, and southeastern Hancock Counties.

J. horizontalis Moench. Creeping Juniper.

Locally rather plentiful. Sandy or rocky banks, often in rock crevices. Restricted essentially to the coast but occasionally inland. Eastern Washington, southern and eastern Hancock, southern Waldo, eastern and southern Knox, southern Lincoln, western Kennebec, southern Sagadahoc, eastern Cumberland, and southeastern York Counties.

J. virginiana L. var. **crebra** Fernald & Griscom. Eastern Red-cedar.

J. virginiana in part. See Rhod. 37:133. 1935.

Frequent and locally plentiful. Dry gravelly or sandy hills, humus-covered rocks and sandy swamps. Central and southern Oxford, northwestern Androscoggin, south central Sagadahoc, Cumberland, and York Counties. Extended by planting to southern Penobscot, north central Knox, southern Lincoln, Kennebec, and western Androscoggin Counties.

Thuja. Arbor-vitae.

T. occidentalis L. Northern White-cedar, Eastern Arbor-vitae.

Common and locally abundant. Chiefly on the basic soils, reaching maximum development in the calcareous regions .(Fernald in Rhodora 21:41-67. 1919). Swamps and cool, rocky banks, sometimes at the higher elevations; also frequently invading old pastures, forming dense stands. Throughout.

LILIACEAE (Lily Family)

Smilax. Green-brier.

S. rotundifolia L. Common Green-brier.

Local, becoming infrequent northward. Moist thickets and sandy or rocky places. Coastal portions of Sagadahoc, Cumberland, and southern York Counties.

SALICACEAE (Willow Family)

Populus. Poplar, Aspen, Cottonwood.

P. alba L. White Poplar.

Native to Europe. Commonly cultivated as an ornamental and frequently spreading widely by the root, often becoming established in waste places, occasionally spontaneous. Planted in the urban centers throughout.

P. balsamifera L. Eastern Cottonwood.
P. deltoides? See Rhod. 21:101. 1919; Sarg. Man. pp. 135-137; Rehd. Man. ed. 2:81. 1940.

Indigenous to northwestern New England and westward. Introduced and commonly planted along streets in the urban centers, but rarely escaping. Throughout. The nomenclature of this and related entities is not wholly clear.

P. grandidentata Michx. Big- or Large-tooth Aspen, Poplar.

Common and locally plentiful. Rich woods, borders of fields and streams. Frequently invading exposed mineral soils, especially after fires. Persisting in deep woods in competition with other hardwoods longer than does the Quaking Aspen. Throughout.

P. nigra L. var. **italica** Muenchh. Lombardy Poplar.

Native to Europe. Introduced commonly as an ornamental. Planted in the urban centers throughout.

P. tacamahacca Mill. Balm-of-Gilead Poplar.
P. candicans. See Rhod. 21:101. 1919.

Origin unknown, perhaps Asiatic. Introduced and frequently planted near the coast and occasionally inland. Northern Aroostook, southern Penobscot, southern Hancock, southern Waldo, Knox, west central Lincoln, southern Sagadahoc, eastern Cumberland, and eastern York Counties.

P. tacamahacca Mill. var. **lanceolata** (Marsh.) Farwell. Balsam Poplar.
P. balsamifera. See Rhod. 21:101. 1919.

Common and widely distributed. Borders of streams and swamps; often in deep low woods where it attains a rather large size. Throughout, but apparently absent from York County.

Fossil leaves of this species have been found in the blue clays of the late Pleistocene of Maine (Tree Ancestors by E. W. Berry, pp. 92-94, Williams & Wilkins Co., Baltimore, 1923).

P. tacamahacca Mill. var. **Michauxii** (Henry) Farwell.
See Rhod. 21:101. 1919.

Sporadic and local. With the species. Northern Aroostook, southeastern Piscataquis, west central Somerset, south central Franklin, and southeastern Hancock Counties.

P. tremuloides Michx. Quaking, American, Trembling, or Small-tooth Aspen; Popple; Aspen.

Common and locally abundant, especially on old burns and in the forests where not over-topped by more dominant species. Light soils chiefly along the borders of woods and other open places; often persisting on rocks at rather high elevations in the mountains. Throughout.

P. tremuloides Michx. var. **magnifica** Marie-Victorin.
See Contrib. du Lab. de Bot. de l'Univ. de Montréal No. 16:10-11. 1930.

Local. With the species. Fort Fairfield (Fernald No. 1692), Aroostook County and Orono (Fernald No. 1697), Penobscot County.

P. tremuloides Michx. var. **pendula** Jaeg.
See Rehd. Man. ed. 2:74. 1940.

Local. Steep clay banks below Eastern Promenade, Portland (Fernald, Long & Norton No. 13407), Cumberland County.

Salix. Willow.

A difficult taxonomic group in which apparently much hybridization occurs. Following the suggestion of Dr. C. R. Ball, certain named species hybrids have been omitted from this list in those cases where the flowering dates of the supposed parents are so widely separated as to make such a cross unlikely.

S. alba L. White Willow.

Native to Europe, and northern Africa to central Asia. Introduced and planted as an ornamental. Occasionally escaping and becoming established near habitation. Northeastern Aroostook, southern Penobscot, east central Hancock, north central Knox, Cumberland, and southern York Counties.

S. alba × fragilis.

Native to Europe and western Asia. Introduced as an ornamental. Occasionally escaped and established principally along the water courses near dwellings. Southern Somerset, southeastern Hancock, northeastern Kennebec, and northeastern York Counties.

S. alba L. var. **calva** G. F. W. Mey. Cricket-bat Willow.

S. alba var. *coerulea*. See Rehd. Man. ed. 2:95. 1940.

Native to Europe and northern Africa to central Asia. Introduced and occasionally planted as an ornamental. Escaped and established along roadsides near habitation in northeastern and southern Washington County.

S. alba L. var. **sericea** Gaud.

See Rehd. Man. ed. 2:95. 1940.

Native to Europe and northern Africa to central Asia. Introduced and occasionally planted as an ornamental but apparently not escaping. Skowhegan, Somerset County.

S. alba L. var. **vitellina** (L.) Stokes. Golden Willow.

Native to Europe and northern Africa to central Asia. Introduced and commonly planted as an ornamental. Spreading freely to roadsides, banks of streams, and waste places near habitation. Established locally in northern Aroostook, southern Piscataquis, and northern Washington Counties, and throughout, in the area south of latitude 45, except Washington, Waldo, and Lincoln Counties.

S. alba var. **vitellina × fragilis.**

Native to Europe and western Asia. Introduced and occasionally cultivated as an ornamental south of latitude 45. Escaped and established principally along the water courses near dwellings. Southern Penobscot, central Waldo, western Knox, eastern Lincoln, western Kennebec, and northeastern Cumberland Counties.

S. arctophila Cockerell.

See Nat. Mus. Canada Bull. No. 92, pt. 1:159-160. 1940.

Local. Moist, sphagnous banks of the wall above Klondike Pond at approximately 3700 feet, Mt. Katahdin (S. Judson Ewer No. 94), Piscataquis County. (See Rhodora 32:260. 1930.)

S. argyrocarpa Anderss.

Local. Mt. Katahdin (Harold J. Dyer, 1940), Piscataquis County. (See also Rhodora 3:172. 1901.)

S. babylonica L. Weeping Willow.

Native to China. Introduced and occasionally cultivated as an ornamental in the urban centers but apparently not escaping. Planted in southeastern Oxford, southeastern Hancock, central Waldo, western Androscoggin, southeastern Cumberland, and central York Counties.

S. Bebbiana Sarg. Bebb Willow.
S. rostrata. See Rehd. Man. ed. 2:102. 1940.

Abundant. Moist or dry ground, sometimes at high elevations in the mountains. Throughout.

S. Bebbiana × humilis.

Local. With the parents. Aziscoos Mountain (B. L. Robinson, July 31, 1903), Oxford County.

S. Bebbiana × petiolaris.

Local. With the parents. Boggy meadow. Pembroke (Fernald No. 1687), Washington County.

S. candida Fluegge. Sage or Hoary Willow.

Rare and local. Cold bogs. Chapman (Hyland No. 875), Woodland (Glen D. Chamberlain No. 1590), and Green Ridge Bog, Caribou (Glen D. Chamberlain No. 1675), Aroostook County.

S. candida × pellita.

Locally abundant on swampy stream bank. In the general vicinity of the parents. Presque Isle (Hyland No. 887), Aroostook County.

S. coactilis Fernald.

Local and infrequent (rather plentiful in southern Penobscot County). Chiefly along the muddy banks of streams and ponds. Northern and northeastern Aroostook, southern Penobscot, east central Piscataquis, south central Somerset, southeastern Waldo, eastern (report) and west central Knox, southeastern Kennebec, west central Sagadahoc (report), and southern York Counties.

S. coactilis × cordata.

Local. With the parents. Orono (Fernald, August 8, 1908) and Pushaw Pond, Orono (Hyland No. 234), Penobscot County.

S. cordata Muhlenb.

Common to abundant. Borders of wet meadows and swamps, and along the water courses generally; depauperate plants often persisting in rocky or gravelly stream beds. Throughout, except Lincoln County.

S. cordata × petiolaris.

Local. With the parents. Swampy thicket below cemetery, Orono (Fernald No. 13364), Penobscot County and Woodmans Mills (Hyland No. 801), Waldo County.

S. cordata × sericea.

Infrequent and widespread. With the parents. Northern and east central Somerset, southern Franklin, northwestern Oxford, east central Washington, and southern York Counties.

S. discolor Muhlenb. Pussy or Glaucous Willow.

Common to abundant. Low meadows, borders of swamps, and wet places along the water courses generally; also frequent on sand plains and other dry places. Throughout.

S. discolor × humilis.

Infrequent. With the parents. Northern Aroostook, east central Penobscot, central Somerset, southeastern Hancock, southern Knox, and southern York Counties.

S. discolor Muhlenb. var. **latifolia** Anderss. Silver Pussy Willow.
S. discolor var. *eriocephala.* See Rehd. Man. ed. 2:99. 1940.

Infrequent. With the species. Southern Penobscot, west central Franklin, southeastern Washington, northwestern Knox, and southern York Counties.

S. discolor Muhlenb. var. **Overi** Ball.

See Rhod. 26:137-138. 1924.

Local. With the species. Southern Penobscot, southeastern Washington, west central Hancock, and northwestern Knox Counties.

S. Elaeagnos Scop.

See Rehd. Man. ed. 2:109. 1940.

Native to central and southern Europe and Asia Minor. Occasionally introduced as an ornamental and locally escaped near dwellings in southwestern Hancock and central Knox Counties.

S. fragilis L. Crack or Brittle Willow.

Native to Europe and western Asia. Commonly introduced as an ornamental, often escaping and becoming established in wet meadows and along the water courses near dwellings. Established locally throughout, south of latitude 45.5. Resistant to the attack of willow blight.

S. glaucophylloides Fernald.

S. glaucophylla. See Rhod. 16:173-175. 1914.

Locally plentiful on river beaches and shores of marly ponds in calcareous districts. Fort Kent and Fort Fairfield, Aroostook County.

S. herbacea L. Dwarf Willow.

Local. In moss or in humus-filled crevices and depressions near the higher mountain summits. Mt. Katahdin, Piscataquis County.

S. humilis Marsh. Prairie Willow.

Common to abundant. Dry plains and barrens and in open, rocky woods; also occasionally in swales and other moist or wet places. Throughout.

S. humilis Marsh. var. **grandifolia** Anderss.

See Anderss. in DC. Prod. 16, pt. 2:236. 1864.

Local. With the species. Fort Kent (Fernald No. 2460), Aroostook County, Bangor (O. W. Knight, May 7 and June 2, 1904), Penobscot County, and Cutler (Kennedy, Williams, Collins & Fernald, July 7, 1902), Washington County. *S. humilis* var. *keweenawensis* (Mich. Acad. Sci. Ann. Rept. 6:206. 1904) is here considered synonymous.

Two forms of this variety have been recognized by Rand and Redfield in the Flora of Mt. Desert Island, Maine, p. 146, 1894: (1) forma *obtusifolia,* near Great Pond, Mt. Desert Island and Sutton Island, Cranberry Isles, Hancock County; and (2) forma *acuminata,* Seal Harbor and Denning Pond, Mt. Desert Island, Hancock County. According to Ball, both forms occur on the same plant.

S. interior Rowlee. Sandbar Willow.

S. longifolia. See Bull. Torr. Bot. Club 27:253. 1900.

Local. Gravelly and cobbly river beaches. Caribou and Fort Fairfield, Aroostook County. Specimen in the New England Botanical Club Herbarium labeled *S. longifolia,* Poland Spring, 1896, Androscoggin County is dubious.

S. lucida Muhlenb. Shining Willow.

Common to abundant. Banks of streams and other wet places, often growing in water along the mucky borders of ponds; also frequently persisting as depauperate plants in crevices of rocky stream beds. Throughout.

S. lucida Muhlenb. var. **angustifolia** Anderss.

Local and infrequent. With the species. Northern and southeastern Aroostook, west central Franklin, west central Oxford, east central Washington, southeastern Hancock, and central Knox Counties.

S. lucida Muhlenb. var. **intonsa** Fernald.

Frequent in the north, becoming more locally common than the species. With the species. Aroostook, Penobscot, southern Piscataquis, east central Somerset, northwestern Franklin, southern Washington, Waldo, eastern Knox, and central Cumberland Counties.

S. lucida Muhlenb. var. **macrophylla** Anderss.
See Anderss. in DC. Prod. 16, pt. 2:205. 1864.

Local. Fort Kent (Williams & Collins, July 22, 1900 and August 10, 1901), Aroostook County. (See Rhodora 3:277. 1901.)

S. nigra Marsh. Black Willow.

Frequent, becoming infrequent northward. Locally plentiful on wet strands, beaches, and wet banks of streams and ponds, often growing in water. Throughout, south of latitude 45, except Washington, northern Hancock, Waldo, and Knox Counties.

A variety [*S. nigra* Marsh. var. *falcata* (Pursh) Torr.] is recognized by some botanists but is not here considered distinct. According to Ball, both forms of the leaf occur on the same plant.

S. pedicellaris Pursh. Bog Willow.

Local. Wet meadows, arbor-vitae swamps, and boggy margins of streams. Northwestern Aroostook, southeastern Penobscot, northern Somerset, and northwestern Oxford Counties.

S. pedicellaris Pursh var. **hypoglauca** Fernald.
See Rhod. 11:161-162. 1909.

Local. With the species and apparently more frequent. Northwestern Aroostook, southern Penobscot, Somerset, east central and southern Oxford, east central Washington, southeastern Hancock, northwestern Androscoggin, and southern Cumberland Counties.

S. pellita Anderss.

Frequent and locally abundant in the north, becoming infrequent southward. Principally confined to bogs, swamps, and banks of streams. Aroostook, east central Penobscot, northern and central Piscataquis, Somerset, northwestern Oxford, southern Kennebec, and southern Sagadahoc Counties.

S. pellita Anderss. forma **psila** Schn.
See Jour. Arnold Arb. 1:83. 1919.

Frequent and locally plentiful. With the species but more often in situations subject to overflow. Western and northern Aroostook County.

S. pentandra L. Bay or Bay-leaved Willow.

Native to Europe. Introduced and commonly cultivated as an ornamental, especially along the coast. Planted in eastern Aroostook, southwestern Piscataquis, east central Washington Counties, and frequently throughout, south of latitude 45. Rarely found as an escape in waste places near dwellings.

Very resistant to the attacks of willow blight.

S. petiolaris Smith.

Common and locally abundant. Damp or wet soil. River thickets and borders of swamps, fields, and meadows, often invading the latter on heavy clay soil. Throughout, except northwestern Aroostook, northern Penobscot, northern and central Piscataquis, northern Somerset, and northern Oxford Counties.

S. petiolaris × sericea.

Local. With the parents. West central Somerset, southwestern Androscoggin, and northeastern Cumberland Counties.

S. petiolaris Smith var. **rosmarinoides** (Anderss.) Schn.
See Jour. Arnold Arb. 2:19. 1920.

Local. With the species. Southern Penobscot, east central Washington, southeastern Hancock, and southern York Counties.

S. planifolia Pursh.
S. phylicifolia. See Rehd. Man. ed. 2:104. 1940.

Local and rare. Mt. Katahdin (Fernald, July 7, 1900), Piscataquis County and Belfast (C. B. Ames in the Boston Society of Natural History Herbarium), Waldo County. (See Churchill in Rhodora 3:155. 1901.) Specimens from Penobscot, Franklin, Waldo, Knox, and York Counties need verification.

S. purpurea L. Purple Willow, Purple Osier.

Native to Europe and northern Africa to central Asia and Japan. Introduced and originally cultivated for basket rods. Now escaped and well established on low ground in waste places near dwellings. Southern Penobscot, southern Piscataquis, southern Somerset, southern Franklin, southeastern Washington, southeastern Hancock, central Waldo, north central Knox, northwestern Androscoggin, and southeastern Cumberland Counties.

S. pyrifolia Anderss. Balsam Willow.
S. balsamifera. See Rhod. 16:116. 1914.

Common and locally plentiful, becoming less frequent southwestward. Swampy woods, damp thickets, and margins of bogs, swamps, and low meadows; also occasionally on wet, rocky slopes in the mountains. Throughout, except York County.

S. sericea Marsh. Silky Willow.

Common and locally abundant. Wet places principally along the water courses on strands and low banks subject to overflow; also along the borders of low meadows and swamps. Throughout, except Lincoln County.

× **S. Smithiana** Willd.
See Rehd. Man. ed. 2:108. 1940.

Native to Europe. Introduced and occasionally planted along roadsides and in hedgerows. Freely spreading and now thoroughly established locally in swampy thickets. Several stations in southeastern Washington County (Rhodora 12:104, 137. 1910).

S. subsericea (Anderss.) Schn.
See Rhod. 11:12, 43. 1909.

Locally abundant in the central portion of the State, becoming infrequent northward and southward. Principally along the borders of wet meadows, grassy bogs, and small streams and ponds; often on heavy, clay soil where it invades the meadows as does *S. petiolaris*. Eastern and central Aroostook, east central and southern Penobscot, southern Piscataquis, central Somerset, southeastern Oxford, east central Washington, eastern Knox, southeastern and northwestern Waldo, northern Lincoln, and southeastern Cumberland Counties.

S. tristis Ait. Dwarf Gray Willow.

A single record known from the State. Harbor Brook, Northeast Harbor (E. L. Rand, August 20, 1889), Hancock County. According to Rand the specimen was collected near "Mr. Eliot's house" and may therefore have been planted. Further evidence that the plant may not be indigenous to this locality is found in the statement: "Rare.... perhaps introduced in this locality from beyond our limits"

(Flora of Mt. Desert Island, Maine, by Rand & Redfield, John Wilson & Son, Cambridge Univ. Press, p. 147. 1894).

S. Uva-ursi Pursh. Bearberry Willow.

Locally rather abundant. Tablelands and slopes near the higher mountain summits. Traveler Mountain (T. 5, R. 9) and Mt. Katahdin (T. 3, R. 9), Piscataquis County and Bald Mountain, Somerset County.

S. viminalis L. Common Osier.

See Rehd. Man. ed. 2:108. 1940.

Native to Europe and northeastern Asia. Introduced and occasionally cultivated as an ornamental, rarely escaping. Planted in Fort Kent, Aroostook County and Bar Harbor, Hancock County.

MYRICACEAE (Sweet Gale Family)

Comptonia (*Myrica*).

C. peregrina (L.) Coulter. Sweet-fern.

Myrica asplenifolia. See Rhod. 40:412. 1938.

Common, becoming sporadic northward; abundant on sandy plains, blueberry barrens, and gravelly hills; also in open dry woods, old fields and on sterile soil generally. Throughout, south of extreme southwestern Aroostook, central Penobscot, southeastern Piscataquis, southern Somerset, southern Franklin, and central Oxford Counties.

Myrica.

M. Gale L. Sweet Gale.

Common and locally abundant. Wet shores, bogs, borders of swamps, and wet depressions in the mountains. Throughout.

M. Gale L. var. **subglabra** (Cheval.) Fernald.

See Rhod. 16:167. 1914.

Frequent. With the species. Southern Penobscot, central and southern Piscataquis, central Somerset, southeastern Oxford, southeastern Washington, western Hancock, northern Knox, central Lincoln, Androscoggin, eastern Cumberland, and southern York Counties.

M. pensylvanica Loisel. Bayberry.

M. carolinensis. See Rhod. 37:423. 1935.

Locally abundant on sandy or sterile soil near the coast; sporadic on the mountains inland. East central and southern Penobscot, southern Washington, southern Hancock, southeastern Waldo, Knox, Lincoln, southern Kennebec, southern Sagadahoc, central and eastern Cumberland, and York Counties.

JUGLANDACEAE (Walnut Family)

Carya (*Hicoria*). Hickory.

C. laciniosa (Michx. f.) Loud. Shellbark or Bigleaf Shagbark Hickory, King Nut.

Indigenous from central New York southwestward. Introduced as an ornamental. Planted locally as a street tree in east central Lincoln County.

C. ovalis (Wangenh.) Sarg. Sweet Pignut.

See Rehd. Man. ed. 2:122. 1940.

Indigenous from Massachusetts westward. Introduced as an ornamental. Planted locally in southeastern Oxford County.

C. ovata (Mill.) K. Koch. Shagbark Hickory.

Rather common in York and southern Cumberland Counties, becoming frequent to infrequent northward. Rich woods and low hills in the vicinity of streams. Southeastern Oxford, western Kennebec, western Androscoggin, Sagadahoc, Cumberland, and York Counties. Locally planted in southern Penobscot, southern Piscataquis, southern Somerset, central Oxford, south central Hancock, east central Waldo, northwestern Knox, and eastern Kennebec Counties.

Juglans. Walnut.

J. cinerea L. Butternut, White Walnut.

Frequent. Rich woods. Southern Somerset, southern Franklin, central and southern Oxford, Kennebec, Androscoggin, southern Sagadahoc, Cumberland, and York Counties. Often cultivated within the borders of its natural range; extended by planting to northern and eastern Aroostook, central Penobscot, southern Piscataquis, central Somerset, northern Oxford, northeastern Washington, Hancock, Waldo, Knox, and Lincoln Counties. Occasionally escaping from cultivation and becoming locally established. The northern boundary of the original range difficult to determine because of specimens originating from fruits gathered from cultivated trees and planted in nearby woods.

J. nigra L. Black Walnut.

Indigenous from western Massachusetts southwestward. Introduced and occasionally planted. Southern Penobscot, southern Somerset, southern Franklin, east central Waldo, west central Knox, southern Sagadahoc, eastern Cumberland, and northwestern and southern York Counties.

BETULACEAE (Birch Family)

Alnus. Alder.

A. crispa (Ait.) Pursh. American Green Alder.

Local. High mountain summits. Mt. Katahdin, Piscataquis County.

A. crispa (Ait.) Pursh var. **mollis** Fernald. Downy Green Alder.
A. mollis. See Rhod. 15:44. 1913.

Common. Exposed rocky banks and damp thickets in the vicinity of streams and ponds; also in the mountains at high elevations. Throughout.

A. incana (L.) Moench. Hoary or Speckled Alder.

Abundant. Swamps and borders of streams and ponds. Throughout.

A. incana (L.) Moench var. **glauca** (Marsh.) Loud.
See Rehd. Man. ed. 2:138. 1940.

Infrequent. With the species. Specimens from Somerset, Penobscot, Oxford, Hancock, Cumberland, and York Counties.

A. incana (L.) Moench var. **glauca** (Marsh.) Loud. forma **tomophylla** Fernald.
See Rhod. 16:56. 1914.

A single station, Hartford (Parlin, August, 1892), Oxford County.

A. incana (L.) Moench var. **hypochlora** Callier.
See Rhod. 23:257. 1921.

Infrequent. With the species. Southern Penobscot, southern Somerset, northern Knox, central Sagadahoc, western Cumberland, and York Counties.

A. rugosa (DuRoi) Spreng. Smooth Alder.

Common in the southwestern part of the State, becoming infrequent northward. Low ground in the vicinity of ponds and streams; often bordering fields and clearings; apparently not ascending to high elevations in the mountains. Stations in all portions of the State south of latitude 45.5.

Betula. Birch.

B. caerulea-grandis Blanchard. Blue Birch.

B. pendula? See Rhod. 24:171-172. 1922.

Infrequent and widely distributed. Rocky upland woods and mountain slopes, usually in exposed places. Northeastern Aroostook, east central Penobscot, east central Piscataquis, south central Somerset, southern Franklin, central Oxford, southeastern Washington, southwestern Hancock, eastern Knox, and western Androscoggin Counties.

B. glandulosa Michx. Dwarf Birch.

Rare. Local on mountain summits. Mt. Katahdin, Piscataquis County.

B. glandulosa Michx. var. **rotundifolia** (Spach) Regel.

Rare. With the species. Mt. Katahdin, Piscataquis County.

B. lenta L. Sweet, Black, or Cherry Birch.

Frequent, becoming infrequent northward. Rich woods, especially in the vicinity of streams. Eastern and southern Oxford, southeastern Hancock (reported as frequent in woods and copses by Rand & Redf., Fl. Mt. Desert Island, Maine, p. 144. 1894 and rare but native at Mt. Desert, by George B. Dorr, The Acadian Forest, Bar Harbor, Maine, 1922), central Kennebec, western Androscoggin, Cumberland, and York Counties.

B. lutea Michx. f. Yellow Birch.

Common forest tree. Rich moist woods, associated with beech and maple on the better soils and with spruce and balsam fir elsewhere. Throughout.

B. lutea Michx. f. var. **macrolepis** Fernald.

See Rhod. 24:170. 1922.

Infrequent. With the species. Southern Piscataquis, southeastern Hancock, east central Kennebec, southwestern Androscoggin, and east central Cumberland Counties.

B. papyrifera Marsh. Paper, White, or Canoe Birch.

B. alba var. *papyrifera.* See Rehd. Man. ed. 2:131. 1940.

Common forest tree. Rich wooded slopes and borders of streams, lakes, and swamps. Scattered through the forests in admixture with other forest trees. Throughout.

B. papyrifera × **populifolia.**

Local. With the parents. Alfred Gore, Alfred (Hyland, July 22, 1936), York County.

B. papyrifera Marsh. var. **cordifolia** (Regel) Fernald. Mountain Paper Birch.

B. alba var. *cordifolia.* See Rehd. Man. ed. 2:131. 1940.

Frequent. Cool woods and mountains, essentially of higher elevations and more exposed habitats than those of the species. Throughout, except Waldo, Lincoln, Kennebec, and Androscoggin Counties.

B. papyrifera Marsh. var.—

B. alba var. *glutinosa.*

Local. With the species. Wassataquoik Valley, Penobscot County and Pembroke, Washington County.

This plant is here tentatively considered as a variety of *B. papyrifera* but further study is necessary in order to determine its correct name and relationships. (See Fernald in The American Journal of Science, fourth series, 14:176-177. 1902.)

B. papyrifera Marsh. var. **minor** (Tuckerm.) Wats. & Coult.
B. alba var. *minor*. See Rehd. Man. ed. 2:131. 1940.

Local. Near the higher mountain summits. Mt. Katahdin, Piscataquis County.

B. populifolia Marsh. Gray, Old-field, or Poverty Birch.

Common, especially near the coast, becoming infrequent northward. Poor sandy or rocky soil, often at high elevations; abundant on old burns, along the borders of old meadows, and other cleared areas; also in damp thickets near swamps and streams. Throughout, except northwestern Aroostook, northern Piscataquis, and northern Somerset Counties.

B. pumila L. Low or Swamp Birch.

Infrequent and local. Open bogs and borders of alder thickets. Central and eastern Aroostook, southern Penobscot, and northern and central Piscataquis Counties.

Carpinus. Hornbeam, Ironwood.

C. caroliniana Walt. var. **virginiana** (Marsh.) Fernald. American Hornbeam, Blue- or Water-beech, Ironwood.
C. caroliniana in part. See Rhod. 37:425. 1935.

Locally common, becoming infrequent northward. Principally restricted to thickets in the immediate vicinity of streams. Southern Penobscot, southern Somerset, southeastern Oxford, southeastern Hancock (report), northern Waldo, western Knox, Lincoln, Kennebec, Cumberland, and York Counties.

Corylus. Hazelnut.

C. americana Walt. American Hazelnut.

Locally common, becoming infrequent northward. Thickets and along streams, especially on the lighter soils. Southern Penobscot, extreme southeastern Somerset, southern Oxford, northwestern Knox, Lincoln, northwestern Kennebec, east central Sagadahoc, Cumberland, and York Counties.

C. cornuta Marsh. Beaked Hazelnut.
C. rostrata. See Rehd. Man. ed. 2:146. 1940.

Common. Open woods, borders of fields, etc., generally on the poorer soils. Throughout.

Ostrya. Hop-hornbeam, Ironwood.

O. virginiana (Mill.) K. Koch. Eastern or American Hop-hornbeam, Ironwood, Leverwood.

Common and locally rather plentiful, becoming less frequent northwestward. Rich woods; also on sparsely wooded, rocky hillsides. Throughout.

FAGACEAE (BEECH FAMILY)

Castanea. Chestnut.

C. dentata (Marsh.) Borkh. Chestnut.

Infrequent and rather sparsely distributed. Indigenous to southern Penobscot (O. W. Knight in Rhodora 8:65-66. 1906), central and southern Oxford, north central Kennebec, Androscoggin, eastern Cumberland, and southern York Counties. Planted and occasionally escaped locally in southeastern Aroostook, southern Penobscot, southern Somerset, central Oxford, and Kennebec Counties.

Barely surviving successive attacks of the chestnut blight.

Fagus. Beech.

F. grandifolia Ehrh. American Beech.

Common forest tree. Abundant in rich upland woods associated with birch and maple. Throughout.

F. sylvatica L. European Beech.

See Rehd. Man. ed. 2:148. 1940.

Native to Europe. Introduced as an ornamental and occasionally planted in the urban centers. Eastern Oxford, central and eastern Hancock, southeastern Waldo, eastern and west central Knox, eastern Cumberland, and eastern York Counties.

F. sylvatica L. var. **atropunicea** West. Purple or Copper Beech.

See Rehd. Man. ed. 2:148. 1940.

Native to Europe. Introduced as an ornamental and frequently planted in the urban centers. Southern Penobscot, eastern Washington, southern Hancock, eastern Waldo, west central Lincoln, western Androscoggin, southern Sagadahoc, southern Cumberland, and southern York Counties.

Quercus. Oak.

Q. alba L. White Oak.

Common and locally abundant, becoming infrequent northward. Dry upland woods and high banks of streams. Southern Franklin, central and southern Oxford, northwestern Knox, Lincoln, Kennebec, Androscoggin, Sagadahoc, Cumberland, and York Counties. Occasionally planted outside its natural range in Portage, Aroostook County and in southern Penobscot, southeastern Somerset, west central Hancock, northern Waldo, and northern Knox Counties. There is some evidence that the Penobscot, Hancock, and Knox stations represent indigenous specimens.

Q. bicolor Willd. Swamp White Oak.

Rather rare and local. Low ground bordering streams and swamps. Northern Androscoggin and eastern York Counties. Planted in southeastern Aroostook, southern Penobscot, southern Somerset, central and southern Oxford, west central Hancock, southwestern Waldo, northwestern Knox, east central Kennebec, northeastern Cumberland, and northwestern York Counties.

Q. borealis Michx. f. Northern Red or Red Oak.

Q. rubra. See Rhod. 18:45-48. 1916; 41:521-524. 1939.

Common and locally abundant, becoming infrequent northward. Dry or moist woods, often in rocky places. Throughout, except western Aroostook, northern Piscataquis, northern Somerset, northern Franklin, and northern Oxford Counties.

Q. borealis × ilicifolia. Lowell Oak.
See Check List, George B. Sudworth, p. 84. U.S.D.A. Misc. Cir. 92. 1927.
A single tree reported by George B. Sudworth (l.c.), Seabury, York, York County.

Q. borealis × velutina. Porter Oak.
See Rehd. Man. ed. 2:160. 1940.
Local. With the parents. Turner (Hyland No. 1202), Androscoggin County, and Alfred (Hyland No. 602), York County.

Q. borealis Michx. f. var. **maxima** (Marsh.) Ashe. Eastern Red or Red Oak.
Q. rubra var. *ambigua.* See Proc. Soc. Am. For. 11:90. 1916.
Frequent. With the species, but apparently sporadic northward. Throughout, except western and southern Aroostook, Piscataquis, northern Somerset, northern and western Franklin, northern Oxford, and northern Washington Counties. Supposedly more common northward than the species. Not separable from the typical form of the species on leaf characters alone.

Q. coccinea Muenchh. Scarlet Oak.
Rare and local. Light soil in rather dry woods. Southwestern Androscoggin, southwestern Cumberland, and central and southern York Counties.

Q. ilicifolia Wang. Bear Oak.
Locally abundant. Sandy barrens and rocky places. Southern Oxford, southeastern Hancock, western Androscoggin, western Sagadahoc, Cumberland, and York Counties.

Q. ilicifolia × velutina.
Local. With the parents. Turner (Hyland No. 1203), Androscoggin County.

Q. macrocarpa Michx. Bur Oak.
Locally plentiful but restricted to the central portion of the State. Rich soil, principally along streams. Central and southern Penobscot, southern Piscataquis, southern Somerset, western Hancock, Waldo, northwestern Knox, and Kennebec Counties. Occasionally planted in eastern Aroostook and throughout its range southward.

Q. montana Willd. Chestnut Oak.
Q. Prinus. See Rehd. Man. ed. 2:172. 1940.
Locally rather plentiful. Dry hillsides, rocky woods and slopes. West central Oxford (report) and southern (mostly in the vicinity of Mt. Agamenticus) York Counties. Somewhat more plentiful than it was a quarter of a century ago, according to A. H. Norton (verbal statement, 1935).

Q. palustris Muenchh. Pin Oak.
Indigenous to Massachusetts and southwestward. Introduced as an ornamental and occasionally planted. Southern Penobscot, southern Somerset, west central and southern Franklin, southeastern Oxford, southeastern Hancock, west central Knox, central Kennebec, eastern Sagadahoc, and northeastern and southern Cumberland Counties.

Q. prinoides Willd. Dwarf Chinquapin Oak.
Dr. Aaron Young (Maine Farmer, May 4, 1848) in a discussion of the species of oaks in the forests of York County enumerates eight species, among them *Q. prinoides.* This species, he states, was found mingled with bear oak on the arid sandy plains of Alfred. From the foliage and the low shrubby habit of the plants Dr. Young at first considered them to be *Q. montana* Willd., but upon finding mature acorns, he finally identified them as *Q. chinquapina* [*Q. prinoides*]. How-

ever, no specimens are available and search in recent years has failed to reveal the plant. (See Norton in Rhodora 37:12. 1935.)

Q. robur L. English Oak.
See Rehd. Man. ed. 2:167-168. 1940.
   Native to Europe, northern Africa and western Asia. Introduced and occasionally planted as an ornamental. Southeastern Hancock, eastern Waldo, southern Kennebec, southwestern Androscoggin, and eastern Cumberland Counties.

Q. velutina Lam. Black Oak.
   Locally abundant, becoming infrequent northward. Barrens and dry or gravelly uplands. Southern Oxford, southern Lincoln, western Androscoggin, Sagadahoc, Cumberland, and York Counties.

ULMACEAE (ELM FAMILY)

Ulmus. Elm.

U. americana L. American or White Elm.
   Common and locally plentiful. Rich, moist woods, especially along the water courses. Often "clannish" when growing in deep woods. Throughout. Much used as an ornamental and street tree in both urban and rural areas.

U. fulva Michx. Slippery or Red Elm.
   Possibly indigenous to the State but certainly now rare. Most of the herbarium specimens labeled *U. fulva* are *U. glabra* Huds. which closely resembles the former. Damp, rich soil. West Farmington (Kate Furbish, May, 1892), Franklin County, Great Deer Island (report), Hancock County, Baldwin (report), Cumberland County, and South Waterboro (E. B. Chamberlain No. 636, July 7, 1898), York County. Planted in Portage, Aroostook County and Limington, York County. Regarding the Hancock station, Norton (Rhodora 37:10-11. 1935) quotes Aaron Young as remarking (in 1848) that *U. fulva* was a native tree of Great Deer Island but that only two or three trees remained standing and these were cruelly stripped of their bark.

U. glabra Huds. Scotch or Wych Elm.
See Rehd. Man. ed. 2:176-177. 1940.
   Native to Europe and western Asia. Introduced and occasionally planted as a street tree. Southern Penobscot, southern Somerset, east central Oxford, central Waldo, eastern Knox, eastern Kennebec, Lincoln, central Sagadahoc, and York Counties.

U. parvifolia Jacq. Chinese Elm.
See Rehd. Man. ed. 2:181-182. 1940.
   Native to China, Korea, and Japan. Introduced and occasionally planted along streets. North central Aroostook, southeastern Washington, and southern Cumberland Counties.

U. pumila L. Siberian Elm.
See Rehd. Man. ed. 2:181. 1940.
   Native to eastern Siberia and northern China. Introduced and occasionally planted along streets. Southern Penobscot, southern Somerset, eastern Hancock, and central Kennebec Counties.

MORACEAE (MULBERRY FAMILY)

Morus. Mulberry.

M. alba L. White Mulberry.
Native to Europe. Introduced and commonly cultivated; often spontaneous near houses. Planted in all sections of the State south of latitude 45, except Washington County.

LORANTHACEAE (MISTLETOE FAMILY)

Arceuthobium. Dwarf Mistletoe.

A. pusillum Peck. Dwarf Mistletoe.
Locally abundant, especially along the coast; sporadic elsewhere. Parasitic on conifers (*Picea* and *Larix*) often causing "witch's brooms" on the host. Throughout, except Waldo, Androscoggin, and York Counties.

RANUNCULACEAE (CROWFOOT FAMILY)

Clematis. Virgin's Bower.

C. verticillaris DC. Purple Clematis.
Infrequent; sporadic and nowhere plentiful. Number of stations now fewer than formerly. Rocky woods and banks of streams, chiefly on calcareous soils. Northern Aroostook, southern Penobscot, central and southern Piscataquis, south central Franklin, west central (report) and southeastern Hancock, west central Knox, southern Kennebec, southeastern Cumberland, and central and southern York Counties.

C. virginiana L. Virgin's Bower.
Common in moist woods and thickets; more abundant along the water courses. Throughout.

MAGNOLIACEAE (MAGNOLIA FAMILY)

Liriodendron. Tulip-tree.

L. Tulipifera L. Yellow-poplar, Tulip-tree, White-wood.
Indigenous from Massachusetts southwestward. Introduced and occasionally planted as an ornamental. Southeastern Aroostook, southern Franklin, eastern Oxford, southeastern Hancock, east central Waldo, central Knox, south central Kennebec, south central Androscoggin, east central Sagadahoc, southeastern Cumberland, and central York Counties.

Magnolia.

M. acuminata L. Cucumber-tree or Cucumber Magnolia.
Indigenous from western New York southwestward. Introduced as an ornamental in the section of the State south of latitude 45. Planted in southern Penobscot, southern Franklin, central Knox, west central Lincoln, south central Androscoggin, Cumberland, and York Counties.

BERBERIDACEAE (BARBERRY FAMILY)

Berberis. Barberry.

B. Thunbergii DC. Japanese Barberry.
See Rehd. Man. ed. 2:237. 1940.

Native to Japan. Introduced and much planted as an ornamental in both urban and rural areas throughout. Escaped and established locally in an old pasture in Rockport (Hyland, June 29, 1934), Knox County, and perhaps occasionally elsewhere.

B. vulgaris L. Common Barberry.

Native to Europe. Introduced; early escaped from cultivation and now thoroughly established in many sections of the State, often distant from habitation. Appearing as indigenous in thickets, sparsely wooded areas, and waste places. Southeastern Aroostook, southern Piscataquis, and all sections of the State south of latitude 45, except Washington County.

LAURACEAE (LAUREL FAMILY)

Lindera (*Benzoin*). Wild Allspice, Fever-bush.

L. Benzoin (L.) Blume. Spice-bush.
Benzoin aestivale. See Rehd. Man. ed. 2:259. 1940.

Occasional and local. Damp woods and thickets on low ground. North central and southern York County. (See Perkins in Rhodora 40:402. 1938.)

Sassafras.

S. albidum (Nutt.) Nees. Sassafras.
S. variifolium in part. See Rhod. 38:178-179. 1936.

Local. Rich woods on the lighter soils. Southern York County.

S. albidum (Nutt.) Nees var. **molle** (Raf.) Fernald. Sassafras.
S. variifolium in part. See Rhod. 38:178-179. 1936.

Local, becoming infrequent northward. Rich woods on the lighter soils. Southern Oxford, eastern Cumberland, and southern York Counties. Planted in western Cumberland and occasionally elsewhere.

SAXIFRAGACEAE (SAXIFRAGE FAMILY)

Ribes. Currant, Gooseberry.

R. americanum Mill. Wild Black Currant.
R. floridum. See Rehd. Man. ed. 2:301. 1940.

Infrequent, becoming frequent northward. Apparently more abundant on soils of calcareous origin. Thickets and rich, moist woods; also spreading along roadsides, especially in eastern Aroostook County. Eastern Aroostook, Penobscot, southern Piscataquis, east central Oxford, east central Washington, southeastern Hancock, Knox, southern Lincoln, Kennebec, southern Cumberland, and southern York Counties.

R. Cynosbati L. Prickly Gooseberry.

Local and infrequent. Rocky woods. West central Somerset (report), southeastern Franklin, central Oxford, and northwestern York Counties.

R. glandulosum Weber. Skunk Currant.
R. prostratum. See Rehd. Man. ed. 2:302. 1940.
 Common. Damp or wet woods and rocks. Throughout.
R. Grossularia L. var. **uva-crispa** (L.) Jancz. European Gooseberry.
See Rehd. Man. ed. 2:308. 1940.
 Native to Europe. Occasionally escaped from cultivation and locally established. East central Oxford County.
R. hirtellum Michx. Smooth Gooseberry.
R. oxyacanthoides. See Coville & Britton North Am. Fl. 22:223-225. 1908.
 Common, becoming somewhat less frequent northwestward. Swamps, grassy swales, and other wet places; also in rocky woods. Throughout.
R. hirtellum Michx. var. **calcicola** Fernald.
See Rhod. 13:76. 1911.
 Infrequent but locally plentiful. Marly swamps and limestone rocks. East central Aroostook, east central and southeastern Penobscot, south central Somerset, southern Franklin, southern Oxford, southern Washington, southern Hancock, eastern and southern Knox, and southern Lincoln Counties.
R. hirtellum Michx. var. **saxosum** (Hook.) Fernald.
See Rhod. 13:76. 1911.
 Local. Mt. Desert Island (Fernald, September 1, 1906), Hancock County.
R. lacustre (Pers.) Poir. Swamp Black Currant.
 Frequent, but mostly absent from the counties along the coast. Cold woods and swamps. Throughout, except Washington, northern and central Hancock, Waldo, western Knox, Lincoln, Kennebec, southeastern Androscoggin, Sagadahoc, central and eastern Cumberland, and York Counties.
R. nigrum L. European Black Currant.
 Native to Europe. Introduced and commonly cultivated for its fruits. Occasionally escaping to roadside thickets and waste places near habitation. Southern Penobscot, southern Somerset, and southeastern Hancock Counties.
R. triste Pall. Swamp Red Currant.
 Infrequent, becoming frequent northward. Mostly on soils of calcareous origin. Cold woods, swamps, and cool, wet depressions in the mountains at rather high elevations. Aroostook, southern Penobscot, southern Piscataquis, northern and central Somerset, west central Oxford, and eastern (report) Washington Counties.
R. triste Pall. var. **albinervium** (Michx.) Fernald.
 Frequent. With the species and apparently somewhat more common. Aroostook, southern Piscataquis, northern Somerset, west central Oxford, and east central Washington Counties.
R. vulgare Lam. Garden Red Currant.
 Native to Europe. Introduced and commonly cultivated for its fruits. Frequently escaping to open woods, fencerows, and waste places. East central Aroostook, southern Penobscot, southern Piscataquis, southern Franklin, east central Oxford, eastern Washington, southeastern Hancock, Knox, Kennebec, northern Androscoggin, Sagadahoc, eastern Cumberland, and eastern York Counties.

HAMAMELIDACEAE (Witch-hazel Family)

Hamamelis. Witch-hazel.

H. virginiana L. Witch-hazel.
 Common and locally plentiful. Damp or dry, often rocky, woods. Throughout,

except the northern portions of Aroostook, Piscataquis, Somerset, Franklin, and Oxford Counties.

H. virginiana L. var. **parvifolia** (Nutt.) T. & G.

See Torr. & Gray Fl. 1:597. 1840; also Rhod. 23:265-266. 1921.

Local. With the species. Southern Penobscot, central Piscataquis, western Androscoggin, eastern Cumberland, and east central York Counties.

PLATANACEAE (PLANE TREE FAMILY)

Platanus. Sycamore, Buttonwood.

P. occidentalis L. American Sycamore, American Plane.

Apparently indigenous but certainly now rare and local. Rich soil, mostly along streams. Southern Oxford (possibly), Kennebec, Cumberland (possibly), and York Counties. Planted occasionally as an ornamental throughout its range. Difficult, from the reports and records available, to distinguish between the planted and indigenous specimens. Reported at an early date by George B. Emerson "along the coast in York County" (Trees and Shrubs of Massachusetts, ed. 3, 1:266. 1878).

ROSACEAE (ROSE FAMILY)

Amelanchier.

Serviceberry, Shad-bush, June-berry, Sugar-plum.

A difficult taxonomic group in which apparently much hybridization occurs. Positive identification of sterile specimens is difficult and records of specific entities are unreliable, especially since the recent recognition of a new species (*A. Wiegandii* Nielsen) which has previously been included with other species.

A. arborea (Michx. f.) Fernald. Downy Serviceberry.

A. canadensis var. *Botryapium*. See Rhod. 43:559-567. 1941.

Infrequent. Gravelly and neutral sandy soils; not a plant of acid sands and granitic soils. Hillsides and dry, open woods. Fort Fairfield, Aroostook, central and southern Penobscot, southern Oxford, eastern Washington, southeastern Hancock, southern Knox, Kennebec, western Androscoggin, Cumberland, and York Counties.

A. canadensis (L.) Medic.

A. oblongifolia in part? See Rhod. 43:559-567. 1941.

Rather common in York County and along the coast, becoming sparse northward. Cold swamps, damp or rocky shores, and wooded mountain slopes; also on sandy plains. In the portions of the State south of latitude 44.5, except Waldo, Lincoln, and Sagadahoc Counties.

A. canadensis × **laevis.**

Rather common, especially in York County. With the parents. Southeastern Penobscot and from Hancock along the coast to York in the counties bordering the sea, except Lincoln and Sagadahoc Counties.

A. canadensis × **stolonifera.**

Infrequent. With the parents. Southeastern Hancock, west central Sagadahoc, and southeastern York Counties.

A. Bartramiana (Tausch) Roem.

A. oligocarpa. See Rhod. 14:158-161. 1912.

Common; locally abundant in cold swamps, damp woods, moist uplands, and on mountain slopes; also in the dry heaths, especially in Washington County. Throughout, except Knox, Lincoln, and Sagadahoc Counties.

A. Bartramiana × laevis.

Frequent. With the parents. Throughout, except Lincoln, Sagadahoc, and York Counties.

A. Bartramiana × sanguinea.

Local. With the parents. Island Falls, Aroostook County and Canada Falls, west branch Penobscot River, Somerset County.

A. Bartramiana × stolonifera.

Infrequent. With the parents. Northern Penobscot, eastern Washington, and southeastern Hancock Counties.

A. Bartramiana × Wiegandii.

Infrequent. With the parents. Aroostook and Washington Counties.

A. gaspensis (Wieg.) Fernald & Weatherby.
See Rhod. 33:235. 1931.

Infrequent. Along the streams and borders of woods and fields. Eastern portion of Aroostook County between Fort Kent and Linneus.

A. gaspensis × laevis.

Local. In the general vicinity of the parents. Oakfield, Aroostook County.

A. gaspensis × Wiegandii.

Infrequent. With the parents. Northeastern Aroostook County in the area east of Fort Kent, Eagle Lake, and Presque Isle.

A. laevis Wieg. Allegheny Serviceberry.
A. canadensis. See Rhod. 14:154-155. 1912.

Common. Chiefly a plant of acid sands and granitic soils. Open damp or dry woods, rocky shores, banks of streams, borders of fields, and other clearings. Throughout, but becoming somewhat less frequent northward.

A. laevis × sanguinea.

Infrequent and widespread. With the parents. Aroostook, east central Piscataquis, extreme northern and extreme southern Somerset, west central Oxford, and west central Hancock Counties.

A. laevis × stolonifera.

Infrequent. With the parents. West central Oxford, southeastern Hancock, eastern Knox, and Cumberland Counties.

A. sanguinea (Pursh) DC.
A. spicata. See Rhod. 14:138-141. 1912.

Frequent but more common in the central and west central part of the State, sporadic elsewhere; mostly absent from the coast. Principally restricted to the gravelly banks of streams. Throughout, except Franklin, Hancock, Knox, Lincoln, Androscoggin, Sagadahoc, and York Counties.

A. stolonifera Wieg.
A. oblongifolia in part. See Rhod. 14:144-147. 1912.

Common, becoming sparse in the heavily wooded portions northward. Sandy barrens, gravelly banks, ledges, and rocky outcrops. Throughout.

A. stolonifera × Wiegandii.

Local. With the parents. Dry heaths of eastern Washington County.

A. Wiegandii Nielsen.

See Am. Midland Nat. 22:160-206. 1939.

Common along the coast, becoming sparse and sporadic inland. Open rocky woods and shores. Locally abundant at Cape Jellison, Stockton Springs, Waldo County. Northeastern Aroostook, southern Penobscot, southern Somerset, Washington (along the coast), Hancock, Waldo, Knox, and southern Sagadahoc Counties.

Crataegus. Hawthorn, Thorn-apple, Thorn.

The treatment of this genus follows that of E. J. Palmer in the Flora of Vermont, ed. 3 rev. 1937, supplemented by suggestions through correspondence (1942). Because of incomplete specimens, insufficient collections and variance in interpretation of the group generally, the following treatment is not considered entirely satisfactory. Since several of the herbarium specimens (especially those of private collections) were not verified by Palmer it is altogether possible that some were incorrectly identified.

C. anomala Sarg.

Local. Orono, Penobscot County.

C. basilica Beadle.

C. alnorum. See Fl. Vermont, ed. 3 rev., p. 151. 1937.

Local. In sandy soil bordering alder thickets. Valley of middle Penobscot River, Orono, Penobscot County. Apparently the specimens collected by Fernald in 1901 (Rhodora 5:153. 1903) represent the sole Maine station as now known. Fernald (l.c.) comments: "The most common species in the region often covering acres of ground with many hundreds of plants."

C. basilica Beadle var. **viridimontana** (Sarg.) Palmer.

See Fl. Vermont, ed. 3 rev., p. 151. 1937.

Local. Primarily a woods species, growing in the dense shade of other trees, but also in moist river thickets. Orono, Penobscot County and Foreside, Cumberland County.

C. Brainerdi Sarg.

Local. Central and southern Penobscot, eastern Hancock, and southern Lincoln Counties.

C. Brainerdi Sarg. var. **asperifolia** (Sarg.) Eggleston.

See Fl. Vermont, ed. 3 rev., p. 151. 1937.

Frequent, becoming sporadic northward. Borders of streams and woods. Eastern Aroostook, southern Penobscot, west central Knox, central Lincoln, central and southern Sagadahoc, Cumberland, and York Counties.

C. Brainerdi Sarg. var. **Egglestoni** (Sarg.) Robinson.

Local. Northeastern Aroostook, southern Piscataquis, and eastern Washington Counties.

C. Brunetiana Sarg.

See Rhod. 5:164-165. 1903.

Local. Northeastern Aroostook and eastern Washington Counties.

C. chrysocarpa Ashe.

See Fl. Vermont, ed. 3 rev., p. 152. 1937.

Infrequent and widely distributed. Northeastern and southern Aroostook, eastern and southern Penobscot, west central Piscataquis, south central Somerset, central Hancock, northeastern Cumberland, and southern York Counties.

C. chrysocarpa Ashe var. **phoenicea** Palmer.
C. rotundifolia. See Fl. Vermont, ed. 3 rev., p. 152. 1937.
Frequent and widely distributed. Throughout, except Somerset, Lincoln, and Kennebec Counties.

C. columbiana Howell var. **Brunetiana** (Sarg.) Eggleston.
Local. Northeastern Aroostook, eastern Washington, southeastern Hancock, southwestern Lincoln, and eastern Cumberland Counties.

C. crus-galli L. Cockspur Thorn.
Indigenous to western New England and regions southwest. Introduced and planted as an ornamental in urban centers. Southern Penobscot, southern Somerset, eastern Hancock, central and southern Waldo, southwestern Androscoggin, eastern Cumberland, and eastern and southern York Counties.

C. dumicola Sarg.
See Rhod. 5: 183. 1903.
Local. River thickets, according to Fernald (l.c.). Northeastern Aroostook and eastern Knox Counties.

C. Faxoni Sarg.
C. rotundifolia var. *Faxoni.* See Rhod. 5:161. 1903.
Local and infrequent. Northeastern Aroostook, southern Penobscot, and northeastern Kennebec Counties.

C. Fernaldi Sarg.
See Rhod. 5:166-167. 1903.
Local. River banks and valleys. Northeastern and southern Aroostook, northern and southern Penobscot, and eastern Knox Counties.

C. fertilis Sarg.
See Rhod. 5:182-183. 1903.
Local and infrequent. River banks in rich alluvial soil. Southern Penobscot, west central Knox, central Lincoln, and northeastern Cumberland Counties.

C. flabellata (Spach) Kirchn.
C. crudelis. See Rehd. Man. ed. 2:362. 1940.
Local. Bank of Penobscot River, Greenbush, Penobscot County. The specimen collected by Hyland (No. 830) in 1937 apparently represents the sole Maine station as now known.

C. Grayana Eggleston.
Local. East central Piscataquis County.

C. Homesiana Ashe.
Local. West central Hancock, southern Kennebec, southeastern Cumberland, and southern York Counties.

C. Jonesae Sarg.
Rather common along the coast; local elsewhere. Southern Somerset, eastern Washington, southern Hancock, southern Waldo, Knox, southeastern Cumberland, and eastern York Counties. One of the most characteristic species of the genus.

C. Keepii Sarg.
See Rhod. 5:165-166. 1903.
Local in river thickets. A common species according to Fernald (l.c.). Fort Fairfield, Aroostook County.
Perhaps not distinct from *C. Brunetiana* Sarg.

C. macrosperma Ashe.
Frequent and locally plentiful. Throughout, south of latitude 46.

C. macrosperma Ashe var. **acutiloba** (Sarg.) Eggleston.
Frequent. Southern Penobscot, southern Piscataquis, southern Somerset, southern Franklin, eastern and southern Oxford, eastern Washington, southeastern Hancock, central Knox, Cumberland, and southern York Counties.

C. macrosperma Ashe var. **matura** (Sarg.) Eggleston.
Local. Southern Aroostook and southern Franklin Counties.

C. monogyna Jacq. English Hawthorn.
See Fl. Vermont, ed. 3 rev., p. 154. 1937.
Native to Europe, northern Africa, and western Asia. Introduced and sparingly escaped from plantings, especially along the coast. Southeastern Washington, southwestern Hancock, southern Kennebec, eastern Cumberland, and southern York Counties.

C. Randiana Sarg.
See Rhod. 5:142-143. 1903.
Local. Along the Aroostook River, Fort Fairfield, Aroostook County and in low moist meadows near Bar Harbor, Mt. Desert Island, Hancock County.
Perhaps not distinct from *C. Brainerdi* Sarg. var. *asperifolia* (Sarg.) Eggleston.

C. submollis Sarg.
Infrequent. Southern Penobscot, southern Piscataquis, southern Somerset, southeastern Cumberland, and southern York Counties.

C. succulenta Schrad. var. **macracantha** (Lodd.) Eggleston.
C. macracantha var. *succulenta*. See Fl. Vermont, ed. 3 rev., p. 155. 1937.
Frequent. Northeastern Aroostook, southern Penobscot, southern Piscataquis, central Hancock, north central and southern Kennebec, north central and eastern Cumberland, and southern York Counties. According to Sargent (Rhodora 7:184-185. 1905), one of the most widely distributed New England species.

Physocarpus. Nine-bark.

P. opulifolius (L.) Maxim. Nine-bark.
Apparently indigenous along rocky banks and shores but appearing more frequently as an escape in waste places near dwellings. Southern Penobscot, southern Hancock, and eastern Cumberland Counties. Planted throughout, except northern and central Aroostook, Penobscot, Piscataquis, Somerset, Franklin, and Oxford Counties.

Potentilla. Cinquefoil.

P. fruticosa L. Shrubby Cinquefoil.
Widespread and locally abundant chiefly on soils of calcareous origin. Wet or dry open ground commonly in bogs and along banks of streams, often the predominant shrub of old rocky pastures. Aroostook, Penobscot, Piscataquis, Somerset, Franklin, east central Oxford, southeastern Washington, east central Hancock, western Kennebec, and Androscoggin Counties.

P. tridentata Ait. Three-toothed Cinquefoil.
Locally common. Exposed sandy and gravelly situations; on rocks and in rock crevices at high elevations in the mountains. Throughout.

P. tridentata Ait. forma **aurora** J. E. Graustein.
See Rhod. 33:211. 1931.
Local. With the species. Isle au Haut, Knox County.

P. tridentata Ait. forma **hirsutifolia** Pease.
See Rhod. 16:194-195. 1914.
Infrequent. With the species. Southern Penobscot, southeastern Hancock, north central Knox, and southern Lincoln Counties.

Prunus. Plum, Cherry.

P. avium L. Sweet Cherry, Mazzard.
Native to Eurasia. Introduced as a fruit tree, occasionally escaping from cultivation and forming thickets near dwellings. Southern Hancock, northern Kennebec, southwestern Androscoggin, southern Cumberland, and east central York Counties.

P. depressa Pursh. Sand Cherry.
P. pumila. See Rhod. 25:69-74. 1923.
Locally common and widely distributed, but rare near the coast. Principally restricted to argillaceous ledges, gravelly beaches, and shores. Northern Aroostook, Penobscot, southern Somerset, southern Franklin, west central Oxford, east central Hancock, northern Kennebec, Androscoggin, and northwestern Cumberland Counties.

P. insititia L. Bullace Plum.
Native to Eurasia. Introduced and cultivated for its fruits. Occasionally planted, sometimes escaping and forming thickets along roadsides and in waste places near dwellings. Central and southern Penobscot and southern Lincoln Counties.

P. maritima Marsh. Beach Plum.
Locally plentiful. Restricted principally to sea beaches and dunes. Eastern Knox, southern Sagadahoc (report), southern Cumberland, and western and extreme eastern York Counties. Occasionally cultivated locally for its fruits.

P. nigra Ait. Wild, Canada, Red, or Horse Plum.
Locally abundant and widely distributed. River banks and roadside thickets; appearing more abundant near dwellings as an escape from cultivation. Throughout, except western Aroostook, northern Piscataquis, northern Somerset, Washington, northern Hancock, Lincoln, and Sagadahoc Counties.

P. pensylvanica L.f. Pin, Wild Red, Bird, or Fire Cherry.
Common; locally abundant in recent clearings, especially following fires. Rocky woods, fencerows, and other open places; also occasionally on rocks in the mountains at rather high elevations. Throughout.

P. serotina Ehrh. Black or Rum Cherry.
Common, becoming sparse northwestward. Rich moist woods, hillsides, high banks of streams, fencerows and other open places; also on rocky cliffs along the sea. Throughout, except northern and western Aroostook, northern Piscataquis, northern Somerset, northern Franklin, and northern Oxford Counties.

P. susquehanae Willd. Appalachian Cherry.
P. cuneata. See Rhod. 25:69-74. 1923.
Local. Dry, sandy barrens and sandy shores. Southern Oxford, southwestern Androscoggin, Cumberland, and northern York Counties.

P. virginiana L. Common Choke Cherry.

Common. Generally in rich, rather moist soil. Fencerows, margins of forests, roadsides, and other open places. Throughout.

P. virginiana L. forma **leucocarpa** (Wats.) Svenson.
See Rhod. 31:99. 1929.

Local. With the species. Southern Aroostook, southern Somerset (report), southwestern Franklin, southeastern Oxford, northwestern Knox, northern Kennebec, and western Androscoggin Counties.

Pyrus.

(Including *Aronia, Malus, Sorbus,* and *Sorbaronia* of some authors.)

Chokeberry, Apple, Pear, Mountain-ash.

P. americana (Marsh.) DC. American Mountain-ash, Roundwood.

Common and locally plentiful. Moist or dry, usually rocky, woods and at high elevations in the mountains. Throughout.

P. americana × **melanocarpa.**

Local. With the parents. Caswell Plantation (Hyland No. 958), Aroostook County.

P. arbutifolia (L.) L.f. var. **atropurpurea** (Britt.) Robinson. Purple-fruited Chokeberry.

Common and locally plentiful. Low woods, swamps, bogs, and wet shores. Throughout.

P. aucuparia (L.) Ehrh. European Mountain-ash, Rowan Tree.

Native to Europe. Introduced and extensively cultivated as an ornamental; often escaping and becoming firmly established locally in fencerows and other open places. Throughout, in and near the urban centers, especially in the eastern and southern sections of the State.

P. aucuparia × **arbutifolia** var. **atropurpurea.**

Local. With the parents. Orono (Hyland No. 225) and Bangor (O. W. Knight, June 9, 1905), Penobscot County.

P. baccata L. Siberian Crab.

Native to Eurasia. Cultivated and occasionally locally established as an escape along borders of woods and roadside thickets near habitation. East central Aroostook and southern Penobscot Counties.

P. communis L. Pear.

Native to Europe. Introduced and cultivated for its fruits. Stray seedlings with degenerate fruit occasionally found as escapes in copses or woods near orchards. Planted in southern Penobscot, southeastern Hancock, southern Knox, west central Lincoln, eastern Cumberland, and York Counties.

P. decora (Sarg.) Hyland.
P. sitchensis. See Rhod. 45:28. 1943.

Frequent and locally plentiful. Banks of streams, shores of lakes and ponds, damp rocky uplands, and mountain slopes. Aroostook, Penobscot, Piscataquis, Somerset, Franklin, Oxford, Washington, Hancock, east central Waldo, and eastern Knox Counties.

P. Malus L. Apple.

Native to Europe. Introduced and cultivated extensively for its fruits. Commonly escaping to woods and thickets over wide areas; frequently found in woods far from highways, often marking the sites of old logging camps. Occurring as an escape throughout, except western Aroostook, northern Penobscot, northern and central Piscataquis, northern and eastern Somerset, and Franklin Counties.

P. melanocarpa (Michx.) Willd. Black-fruited Chokeberry.

Common and locally abundant. Moist woods, barrens, rocky uplands, and at high elevations in the mountains. Throughout.

× **P. prunifolia** Willd. Crab Apple.

Native to northeastern Asia. Introduced and cultivated as crab apples. Occasionally escaping to roadsides and thickets near dwellings. Southern Penobscot and west central Hancock Counties.

Rosa. Rose.

R. acicularis Lindl.

Apparently a species inhabiting the region north and west of our borders, but several specimens have been collected along the coast as long ago as 1891. Roadsides, dry pastures, beaches, and shores. Eastern Knox, central Lincoln, southern Sagadahoc, and eastern Cumberland Counties.

R. acicularis Lindl. var. **Bourgeauiana** Crepin.

A characteristically northern rose occurring outside our borders, but several stations have been recorded for the State. Western Aroostook (report), east central Franklin, eastern Knox, southwestern Sagadahoc, eastern Cumberland, and northeastern York Counties.

R. blanda Ait.

Infrequent and local, becoming common northward. Shores and rocky places. Aroostook, east central Penobscot, central and southern Piscataquis, central and southern Somerset, southern Franklin, southeastern Hancock, and northern Kennebec Counties.

R. blanda × **palustris.**

Local. With the parents. Gravelly thickets and shores. Southeastern Penobscot and southern Piscataquis Counties.

R. blanda × **virginiana.**

Local. With the parents. Pleasant Ridge Plantation (J. Franklin Collins, July 24, 1896), Somerset County.

R. blanda Ait. var. **hispida** Farwell.

See Papers Mich. Acad. Sci. 2:25. 1923.

Infrequent. With the species. Southern Penobscot and southern Somerset Counties.

R. carolina L.

R. humilis. See Rhod. 20:91 (footnote). 1918.

Common, becoming sporadic and infrequent northward. Dry soil and rocky places; also on the lower mountain slopes. Northern Aroostook, central Penobscot, west central Somerset Counties, and throughout, south of latitude 45 except southern Franklin, northern Oxford, northwestern Washington, northern Hancock, and western Waldo Counties.

R. carolina × palustris.
Local. With the parents. Southeastern Washington and southern Cumberland Counties.

R. cinnamomea L. Cinnamon Rose.
Native to Eurasia. Introduced and commonly cultivated; often escaped to waste places near habitation. Northeastern Aroostook, southern Penobscot, southern Piscataquis, southeastern Hancock, southwestern Waldo, eastern Knox, northern Androscoggin, Sagadahoc, and east central York Counties.

R. Eglanteria L. Sweet-brier, Eglantine.
R. rubiginosa. See Rehd. Man. ed. 2:435. 1940.
Native to Europe. Introduced and commonly planted as an ornamental. Escaped and widely established in rocky pastures and along roadsides and waste places near habitation. Southeastern Aroostook, southern Penobscot, southern Piscataquis, southern Somerset, Franklin, west central Oxford, southeastern Hancock, southern Waldo, eastern Knox, southern Kennebec, western Androscoggin, southern Sagadahoc, Cumberland, and York Counties.

R. gallica L. French Rose.
Native to Europe. Introduced and often cultivated. Frequently escaped and established along roadsides near habitation. Southeastern Franklin, southern Knox, southern Kennebec, eastern Cumberland, and northwestern York Counties.

R. johannensis Fernald.
See Rhod. 20:90-96. 1918.
Infrequent, becoming locally plentiful northward. Wet, gravelly banks and shores. Northern Aroostook, southern Penobscot, central and southern Somerset, and north central Knox Counties.

R. johannensis × palustris.
Local. With the parents. Roadside thicket, Portage Lake (Robinson & Fernald, August 9, 1901), Aroostook County.

R. micrantha Smith.
R. rubiginosa var. *micrantha.* See Rehd. Man. ed. 2:435. 1940.
Native to Europe. Introduced and occasionally planted as an ornamental. Locally established at North Berwick, York County.

R. nitida Willd.
Frequent. Principally along the margins of swamps and bogs. Throughout, except northeastern Aroostook, northern and central Piscataquis, central and southern Somerset, northern Oxford, Waldo, Lincoln, and western Cumberland Counties.

R. nitida × palustris.
Local. With the parents. Southern Piscataquis, southwestern Somerset, and southern Franklin Counties.

R. nitida × virginiana.
Local. With the parents. Grassy marsh. Jefferson (Hyland No. 112), Lincoln County.

R. palustris Marsh.
R. carolina. See Rhod. 20:91 (footnote). 1918.
Frequent, becoming infrequent and sporadic northward. Margins of swamps and banks of streams. Throughout, except western and southeastern Aroostook, northern Penobscot, northern and western Piscataquis, Somerset, northern Franklin, northern Oxford, northern and western Washington, northern Hancock, and northern Waldo Counties.

R. rugosa Thunb. Rugosa Rose.
See Rehd. Man. ed. 2:438. 1940.
Native to northern China, Korea, and Japan. Introduced and commonly cultivated as an ornamental. Often escaping to fencerows, roadsides, rocky pastures, and shores; also frequent on rocks and sandy places along the coast. Established in east central Aroostook, southern Penobscot, southeastern Hancock, and Knox Counties. Commonly planted in the area of the State south of latitude 45.

R. spinosissima L. Burnett or Scotch Rose.
Native to Eurasia. Introduced and often cultivated, frequently spreading from old gardens especially along the coast. Northeastern Aroostook, south central Somerset, southern Hancock, southern Waldo, Knox, and southeastern York Counties.

R. virginiana Mill.
Common, becoming local and infrequent northward. Essentially in rather dry habitats, but also along rocky shores and margins of swamps. North central Aroostook, central Penobscot, and northern Washington Counties, and throughout, south of latitude 45.

R. virginiana Mill. var. **lamprophylla** Rehd.
See Rehd. Man. ed. 2:437. 1940.
Local. Eastern Hancock and southwestern Androscoggin Counties.

Rubus.

Bramble, Blackberry, Dewberry, Raspberry.

In general the treatment of this genus follows L. H. Bailey in Gentes Herbarum vol. 5 for these groups which have been published to date (through fasc. 7, 1944). However, many species have not yet been worked up for Maine and it is impossible to make a complete authentic list at this time. Following the suggestion of Dr. Bailey, formerly listed hybrids are not considered since they appear to have little meaning in this complex group. Strictly herbaceous species have been omitted from this list.

R. adjacens Fernald.
See Gent. Herb. 5:92. 1941; Rhod. 42:290-293. 1940.
Dry soil, Pembroke, Washington County and border of moist woods and thickets, Lyman and West Kennebunk, Kennebunk, York County.

R. allegheniensis Porter.
Abundant, becoming infrequent northward. Dry, open thickets and recent clearings. Throughout.

R. allegheniensis Porter var. **calycosus** Fernald.
See Gent. Herb. 2:381. 1932; Rhod. 10:51. 1908.
Reported for Maine (Gent. Herb. l.c.) without specific localities.

R. allegheniensis Porter var. **neoscoticus** (Fernald) Bailey.
See Gent. Herb. 2:382-383. 1932; Rhod. 23:268. 1921.
Infrequent. With the species. East central Aroostook, southeastern Penobscot, and southern Sagadahoc Counties.

R. alter Bailey.
See Gent. Herb. 5:82-84. 1941.
Local. Wells (reported by Bailey, l.c.), York County.

R. amicalis Blanchard.
R. canadensis in part. See Gent. Herb. 2:375-376. 1932; 5:476-478. 1944; Rhod. 13:55-56. 1911.

Local. Pastures and roadsides on dry or open land. Mt. Desert and nearby mainland, Hancock County (report) and southeastern York County.

R. arenicolus Blanchard.
See Rhod. 8:151-152. 1906; Gent. Herb. 5:320-322. 1943.

Locally abundant. Open sandy and gravelly places. Southwestern Androscoggin and east central York Counties.

R. arenicolus Blanchard var. **confictus** Bailey.
See Gent. Herb. 5:322. 1943.

Local. Rocky ledges and banks. Old Orchard, York County.

R. argutus Link.

Local and infrequent. Dry, open ground and thin woods. Southern Penobscot, southern Piscataquis, southwestern Knox (report), and southern Lincoln Counties.

R. armatus (Fernald) Bailey.
See Gent. Herb. 5:338. 1943.

Local. St. Francis (Pease & Goodale, July 9, 1932), Aroostook County.

R. arundelanus Blanchard.
R. recurvans. See Gent. Herb. 5:373-376. 1943; Rhod. 8:176-177. 1906.

Locally plentiful. Open places on rocks and in rich or poor soil. East central Knox and eastern York Counties.

R. arundelanus Blanchard var. **Jeckylanus** (Blanchard) Bailey.
R. Jeckylanus. See Gent. Herb. 5:376. 1943.

Local. Open, rocky, and sandy places near the coast. East central York County. (See Gent. Herb. 3:261-263. 1934, for comments by Bailey.)

R. biformispinus Blanchard.

Locally abundant. Open places in dry ground. East central York County.

R. canadensis L.

Common and locally abundant the farthest north of any of the highbush blackberries. Rocky soil in mountains and highlands; also in thickets and cool hollows about lakes and wood margins. Throughout, except Kennebec County.

R. canadensis L. var. **elegantulus** Farw.
See Gent. Herb. 2:374. 1932; Am. Midland Nat. 11:263. 1929.

Occasional. With the species. Rocky and gravelly uplands. East central Piscataquis, east central Penobscot, and east central and southern Oxford Counties.

This entity is no longer kept separate, by Bailey, as a variety of *R. canadensis*.

R. canadensis L. var. **pergratus** (Blanchard) Bailey.
R. pergratus. See Gent. Herb. 5:470. 1944.

Rather abundant, becoming infrequent northward. Meadows, river banks, and moist or dry ground in open sun or light shade. East central Penobscot, southern Piscataquis, and throughout, south of latitude 45, except Washington and Waldo Counties.

R. canadensis L. var. **Randii** Bailey.
R. Randii. See Gent. Herb. 3:261. 1934.

Local. In shady thickets. East central Penobscot, southeastern Hancock (report), and southern Cumberland Counties.

This variety has recently been subjugated by Bailey (Gent. Herb. 5:472. 1944).

R. facetus Bailey.
See Gent. Herb. 2:449. 1932.
   Local. Roadside. Kennebunkport (Hyland No. 308), York County.

R. flagellaris Willd. Common Eastern Dewberry.
R. villosus var. *roribaccus*. See Gent. Herb. 5:244-250. 1943.
   Locally common. Dry, open places. Southern Somerset, northeastern and southwestern Cumberland, and York Counties.

R. frondisentis Blanchard.
   Local. Dry, open places. Fryeburg (Hyland No. 490), Oxford County.

R. geophilus Blanchard.
See Gent. Herb. 5:260-262. 1943; Rhod. 8:148. 1906.
   Locally common. Dry, open places. East central York County.

R. glandicaulis Blanchard.
   Rather common, becoming infrequent northeastward. Dry, open places. Southern Penobscot, southern Piscataquis, east central and southeastern Washington, southeastern Hancock, southeastern Waldo, northern and eastern Knox, eastern Cumberland, and east central York Counties.

R. Groutianus Blanchard.
See Gent. Herb. 5:178-180. 1941.
   Probably widespread. Margins and open mostly sandy grounds, low borders, grassy fields, and along wood roads. Reported for Maine (l.c.) without specific localities.

R. harmonicus Bailey.
See Gent. Herb. 5:86-89. 1941.
   Local. Kennebunk and Wells (reported by Bailey, l.c.), York County.

R. heterophyllus Willd.
R. recurvans? See Gent. Herb. 2:409-412. 1932; Berl. Baumz. ed. 2:413. 1811.
   Locally abundant, becoming infrequent northward. Dry, open places and borders of woods. East central and southeastern Aroostook, central and southern Penobscot, southern Piscataquis, southern Somerset, southeastern Hancock, northeastern Kennebec, northeastern and southwestern Androscoggin, and York Counties.

R. hispidus L. Swamp Dewberry.
   Common and locally plentiful, becoming infrequent northward. Low woods, swampy meadows, bogs, and subacid mossy uplands. Northeastern Washington County and throughout, south of latitude 45.5.

R. hispidus L. var. **obovalis** (Michx.) Fernald.
See Gent. Herb. 5:75. 1941; Rhod. 42:281-282. 1940.
   Frequent. With the species and covering the same territory.

R. idaeus L. Red Raspberry.
   Common and locally abundant. Thickets, clearings, and rich woods. Throughout.

R. idaeus L. var. **aculeatissimus** (C. A. Mey.) Regel & Tiling. Wild Red
   Raspberry.
   Frequent, becoming infrequent northward. Thickets, moist banks, and rich woods. North central Aroostook, southern Penobscot, west central (report) and southern Somerset, southern Franklin, central and southern Oxford, southeastern Washington (report), Hancock, Knox, and southern Cumberland Counties.

R. idaeus L. var. **canadensis** Richardson.
See Rhod. 21:97. 1919.
   Frequent, becoming sparse northward. With the species. North central and east central Aroostook, central Piscataquis, northern Somerset, southwestern Oxford, southeastern Hancock, southern Knox, southern Lincoln, and southern York Counties.

R. idaeus L. var. **canadensis** Richardson forma **Warei** Deane & Fernald.
See Rhod. 22:112. 1920.
   Local. Rich woods. Slopes of Caribou Mountain, Mason (report), Oxford County.

R. idaeus L. var. **heterolasius** Fernald.
See Rhod. 21:97. 1919.
   Local. Steep clay banks and open places in the immediate vicinity of the coast. Southern Lincoln, southern Sagadahoc, and southeastern Cumberland Counties.

R. idaeus L. var. **strigosus** (Michx.) Maxim.
See Rhod. 21:96. 1919.
   Frequent, becoming infrequent northward. With the species. Eastern Aroostook, southern Penobscot, east central and southern Piscataquis, southern Somerset, central Oxford, southeastern Washington, southern Hancock, eastern and southern Knox, southern Lincoln, Androscoggin, southern Sagadahoc, eastern Cumberland, and eastern and southern York Counties.

R. idaeus L. var. **strigosus** (Michx.) Maxim. forma **tonsus** Fernald.
See Rhod. 21:96-97. 1919.
   Local and infrequent. With the species. Alluvial or boggy woods and thickets. Southern Piscataquis, southern Somerset, northeastern Kennebec, northwestern and southern York Counties.

R. junceus Blanchard.
   Local and infrequent. Dry land. Western Aroostook, southern Penobscot, southern Piscataquis, east central Oxford, southeastern Hancock, and east central York Counties.

R. mainensis Bailey.
See Gent. Herb. 5:325. 1943.
   Local. Dry land. Millinocket, Penobscot County.

R. miscix Bailey.
R. peculiaris. See Gent. Herb. 1:257. 1925; 2:360. 1932; 5:478-480. 1944.
   Local and infrequent. Dry ground. Southern Aroostook, southern Piscataquis, eastern and southern Knox, southern Lincoln, and northeastern York Counties.

R. multiformis Blanchard.
See Gent. Herb. 2:377. 1932.
   Local and infrequent. Dry ground in sun or light shade. Southeastern Hancock, eastern Knox, and northeastern York Counties.

R. multilicius Bailey.
See Gent. Herb. 5:480. 1944.
   Local. Open dry land. Near Kidney Pond Camps, T. 3, R. 10 (Bailey No. 835), Piscataquis County.

R. nigricans Rydb.
See Gent. Herb. 5:140-142. 1941.
   Infrequent. Open, dry places. Southern Penobscot, east central and south-

western Oxford, northern Washington, northern Androscoggin, and southern York Counties.

R. occidentalis L. Black Raspberry.

Local and infrequent. Copses and waste places; often cultivated and escaped near dwellings. Southern Penobscot, southern Piscataquis, southern Somerset, central and southern Oxford, northeastern Kennebec, northern and southwestern Androscoggin, west central Sagadahoc, northeastern and northwestern Cumberland, and York Counties.

R. odoratus L. Purple Flowering Raspberry.

Frequent. Rocky woods and copses. Throughout, south of latitude 45, except Washington, Waldo, and Knox Counties. Difficult in some cases to determine from the records whether indigenous or escaped. Commonly cultivated and occasionally established in eastern Aroostook County and in the area south of latitude 45, including Knox County.

Fassett (Ann. Mo. Bot. Gard. 28:299-374. 1941) recognizes four forms of *R. odoratus* occurring in Maine as follows: forma *hypomalacus* Fassett, Augusta, Kennebec County, and Paris, Oxford County; forma *heteradenius* Fassett, Vassalboro and Augusta, Kennebec County, and Paris, Oxford County; forma *bifarius* Fassett, Vassalboro and Augusta, Kennebec County, and Paris, Oxford County; and forma *glabrifolius* Fassett, Vassalboro and Augusta, Kennebec County.

R. ostryifolius Rydb.

R. Andrewsianus. See Gent. Herb. 2:392-393. 1932.

Infrequent. Fields and open thickets. Southern Penobscot, Knox, and southern Lincoln Counties.

R. pergratus Blanchard.

Rather abundant, becoming infrequent northward. Meadows, river banks, and moist or dry ground in open sun or light shade. East central Penobscot, southern Piscataquis, and throughout, south of latitude 45, except Washington and Waldo Counties.

R. permixtus Blanchard.

Local. Dry to moist soil, margins of low areas, and in swales. Biddeford (report), York County. (See Gent. Herb. 3:262. 1934; and 5:118-120. 1941.)

R. plicatifolius Blanchard.

See Gent. Herb. 5:338-340. 1943; Rhod. 8:149. 1906.

Locally abundant. Open places, especially in sandy ground. Southeastern Hancock and southern York Counties.

R. pudens Bailey.

See Gent. Herb. 5:78-80. 1941.

Local. Lee (Steinmetz, July 20, 1937) and Springfield (Hyland No. 1172), Penobscot County, and Ocean Point, Boothbay Harbor (Fassett No. 7268), Lincoln County.

R. recurvicaulis Blanchard.

See Gent. Herb. 5:333-336. 1943; Rhod. 8:153. 1906.

Frequent and locally plentiful, becoming infrequent northward. Roadsides and pastures; also along rivers and streams. Northern and central Aroostook, southern Penobscot, southern Piscataquis, southern Somerset, east central Oxford, southeastern Washington, southern Hancock, Knox, southern Lincoln, northeastern Kennebec, central Androscoggin, northwestern Sagadahoc, and eastern and central York Counties.

R. rhodinsulanus Bailey.
See Gent. Herb. 5:325-327. 1943.
Local. Open, dry places or among bushes on the margins. Biddeford (report), York County. (See Gent. Herb. 3:262. 1934.)

R. Rossbergianus Blanchard.
See Rhod. 9:7. 1907.
Local. Open, dry places. North Berwick (Parlin, June 24, 1893), York County.

R. semisetosus Blanchard var. **ortivus** Bailey.
See Gent. Herb. 3:254. 1934.
Local. Long Pond Meadows, Mt. Desert Island, Hancock County. Reported by Bailey (l.c.).

R. tardatus Blanchard.
Local and frequent, becoming infrequent northward. Sandy or other dry places. Southern Somerset, west central Franklin, southeastern Washington, southeastern Hancock, south central Knox, southern Lincoln, southern Cumberland, and southeastern York Counties.
According to Bailey (Gent. Herb. 5:104. 1941) the reported ranges outside of York County require verification.

R. vermontanus Blanchard.
See Gent. Herb. 5:154-161. 1941; Am. Bot. 7:1. 1904.
Local and frequent. Thickets and open places mostly in dry soil; also in moist woods, swales, and bogs. North central and southern Penobscot, southern Somerset, central Franklin, central Oxford, southern Washington, west central and southeastern Hancock, southern Lincoln, and northern York Counties.

Sorbaria.

S. sorbifolia (L.) A. Br. Ural False-spiraea.
Native to Asia. Commonly cultivated as an ornamental and often escaped to roadsides and waste lands near dwellings. Established near habitation in all sections of the State, except western Aroostook, northern Penobscot, northern and central Piscataquis, northern and central Somerset, northern Franklin, and northern Oxford Counties.

Spiraea. Spiraea.

S. latifolia (Ait.) Borkh. Meadowsweet.
Common and locally abundant. Rocky pastures, barrens, borders of woods, and other open places. Throughout.

S. latifolia (Ait.) Borkh. var. **septentrionalis** Fernald.
See Rhod. 19:254-255. 1917.
Local. Mountainous regions. Depot Pond, Mt. Katahdin (Williams, July 16, 1900), Piscataquis County.

S. tomentosa L. Hardhack, Steeple Bush.
Common and locally abundant. Old fields, hilly pastures, barrens, and on open, low ground generally. Southeastern Aroostook, central Penobscot, southern Piscataquis, south central Somerset, and northern Washington Counties, and throughout, south of latitude 45.

LEGUMINOSAE (Pulse Family)

Baptisia. False Indigo.

B. tinctoria (L.) R. Br. Wild Indigo.

Possibly indigenous to the State, the natural range apparently extending north as far as southern New Hampshire. Local. Alfred, York County. (See Perkins in Rhodora 40:402. 1938.)

Caragana.

C. arborescens Lam. Pea-tree.
See Rehd. Man. ed. 2:514. 1940.

Native to Siberia and Manchuria. Introduced; cultivated and rarely escaped near dwellings. Southern Penobscot, eastern Hancock, and eastern Cumberland Counties.

Cladrastis. Yellow-wood.

C. lutea (Michx. f.) Koch. Yellow-wood.

Indigenous to the southern and east central states. Introduced as an ornamental and occasionally planted in southern Penobscot, southern Somerset, east central Waldo, central Lincoln, east central Kennebec, and eastern Cumberland Counties.

Cytisus. Broom.

C. scoparius (L.) Link. Scotch Broom.

Native to Europe. Introduced and sparingly escaped in sandy places along the beach near habitation. Southeastern York County. (See Rhodora 37:416. 1935.)

Genista. Woad-waxen, Whin.

G. tinctoria L. Dyer's Greenweed.

Native to Europe. Introduced as an ornamental and occasionally escaped near dwellings, becoming established on sterile hills and along roadsides. Southeastern Hancock, eastern Cumberland, and east central and southern York Counties.

Gleditsia. Honey-locust.

G. triacanthos L. Honey-locust.

Indigenous from western New York and Pennsylvania south and westward. Introduced as an ornamental and for hedges, but seldom escaped. Frequently planted in the urban centers throughout, but less commonly northward.

Gymnocladus. Kentucky Coffee-tree.

G. dioicus (L.) K. Koch. Kentucky Coffee-tree.

Indigenous from central New York and Pennsylvania south and westward. Only occasionally introduced as an ornamental. Planted in southern Penobscot, south central Androscoggin, and northwestern York Counties.

Robinia. Locust.

R. hispida L. Bristly Locust, Rose-acacia.

Indigenous in the mountains from Virginia to Georgia. Introduced and occasionally cultivated in southwestern Piscataquis County and in the area south of latitude 45 and west of longitude 69. Escaped and established in the fields at Wellington (fide R. B. Trask), Piscataquis County.

R. Pseudo-Acacia L. Black Locust.

Indigenous in the mountains from Pennsylvania to Georgia and westward. Introduced and commonly cultivated as an ornamental and for its timber. Escaped and thoroughly naturalized over wide areas, principally along sandy banks and near dwellings. Throughout, except in the densely wooded areas of western Aroostook, northern Piscataquis, northern Somerset, Franklin, and northern Oxford Counties.

R. viscosa Vent. Clammy Locust.

Indigenous in the mountains from Virginia to Georgia. Introduced and cultivated as an ornamental. Often escaped and naturalized along roadsides, sandy banks, and waste places near habitation. Eastern Aroostook, southwestern Piscataquis, and northeastern Washington Counties, and throughout, south of latitude 45, except Oxford County.

RUTACEAE (RUE FAMILY)

Ptelea. Shrubby Trefoil, Hop-tree.

P. trifoliata L. Common Hop-tree, Wafer-ash.

Indigenous to New York and westward. Occasionally introduced and cultivated as an ornamental. Planted in southern Penobscot, southern Somerset, north central Knox, southeastern Cumberland, and east central York Counties.

Zanthoxylum (*Xanthoxylum*). Prickly-ash.

Z. americanum Mill. Northern Prickly-ash, Toothache-tree.

Indigenous to the region from western Quebec to Minnesota and southward. Occurring in the State locally at Shaker Hill, Alfred, York County, where it was probably introduced by the Shaker herbalists at an early date (Perkins in Rhodora 40:462-464. 1938).

EMPETRACEAE (CROWBERRY FAMILY)

Corema. Broom Crowberry.

C. Conradii Torr. Broom Crowberry.

Locally rather abundant. Dry, boulder-strewn soils and crevices in rocky outcrops; also on sandy pine barrens. Strictly confined to the islands and coastal region. Southern and eastern Hancock, Knox, southern Lincoln, southern Sagadahoc, and eastern Cumberland Counties.

Empetrum. Crowberry.

E. atropurpureum Fernald & Wieg. Red Crowberry.

E. nigrum var. *andinum*. See Rhod. 15:214-215. 1913.

Infrequent, locally rather abundant. Gravel or granite rocks in the mountains at or near the timber line. East central and southwestern Piscataquis, west central Somerset, west central and southern Franklin, central Oxford, east central Hancock, and west central and southern York Counties.

E. nigrum L. Black Crowberry.

Widely distributed but restricted to the coast and the higher mountains inland.

Wet, peaty meadows and rocky places along the coast and in peat-moss or humus near the summits of the higher mountains inland. East central and southern Piscataquis, west central Somerset, central Franklin, central Oxford, eastern Washington, southern Hancock, eastern Knox, southern Lincoln, southern Sagadahoc, and eastern Cumberland Counties.

ANACARDIACEAE (CASHEW FAMILY)

Rhus. Sumac.

R. copallina L. Dwarf or Shining Sumac.

Infrequent and local. Rocky or rather sandy places. Throughout, south of latitude 44.

R. copallina L. var. **latifolia** Engler.
See Rhod. 37:167-168. 1935.

Infrequent and local. West central Kennebec County and with the species in the area south of latitude 44.

R. glabra L. Smooth Sumac.

Common, becoming infrequent northward. Dry soil principally along roadsides and high banks of streams. Throughout, south of latitude 45, except Washington, Hancock, Waldo, eastern Knox, southern Lincoln, and Sagadahoc Counties. Occasionally planted in eastern Penobscot and southeastern Washington Counties.

R. glabra L. var. **borealis** Britt.
See Britt. Man. Fl. North. States & Can. p. 601. 1901.

Local. With the species. Southern Somerset and southern York Counties.

R. radicans L. Poison Ivy, Poison-oak.
R. Toxicodendron. See Rehd. Man. ed. 2:544. 1940.

Common and locally abundant. Thickets, fencreows, and open woods on a variety of soil types ranging from wet to dry; abundant on rocks, especially near the coast, but seldom at high elevations. Apparently less abundant north and northeastward. Throughout.

Fernald (Rhodora 43:589-596. 1941) has recently published on several varieties and forms of *R. radicans* occurring within the Gray's Manual range. Data on the occurrence and distribution within the State are incomplete but the following are listed from the available records:

R. radicans L. forma **malacotrichocarpa** (A. H. Moore) Fernald.
See Rhod. 11:162-163. 1909; 43:589-596. 1941.

Local (and probably infrequent). Pemaquid Beach, Bristol (Chamberlain & Dinsmore No. 832), Lincoln County.

R. radicans L. var. **Rydbergii** (Small) Rehd.
See Rhod. 43:589-596. 1941; Jour. Arnold Arb. 20:416. 1939.

Common and locally abundant. Woods, rocky slopes and wet peat. Fort Fairfield (Fernald No. 1995), Aroostook County, Winn (Fernald & Long No. 14014), Penobscot County, Pembroke (Fernald No. 1993), Washington County, and Camden (Rossbach No. 532), Knox County.

R. radicans L. var. **vulgaris** (Michx.) DC.
See Rhod. 43:589-596. 1941.

Probably common. Swampy woods. Fairfield (Fernald & Long No. 14017), Somerset County.

Note: *R. radicans, including its forms and varieties, is poisonous to touch.*

R. typhina L. Staghorn Sumac.

Common and locally abundant, becoming infrequent north and northwestward. Dry or gravelly banks and roadsides. Throughout, except western Aroostook, northern Piscataquis, northern Somerset, northern Franklin, and northern Oxford Counties.

R. typhina L. forma **laciniata** (Wood) Rehder.
See Rhod. 9:115. 1907.

Local. Saco (W. H. Manning, July 23, 1886), York County. Occasionally planted as an ornamental in southern Penobscot and western Hancock Counties.

R. vernix L. Poison Sumac, Poison-dogwood.

Frequent, becoming sparse and sporadic northward. Swamps and wet shores. Orono, Penobscot County, Chesterville, Franklin County, Matinicus Island (report), Knox County, Bristol (report), Lincoln County, Leeds, Androscoggin County, Cumberland, and York Counties.

Note: *This species is poisonous to touch.*

AQUIFOLIACEAE (HOLLY FAMILY)

Ilex. Holly.

I. glabra (L.) Gray. Inkberry.

Rare and local. Bogs and low wet or dry sandy soil. Isle au Haut, Knox County.

I. laevigata (Pursh) Gray. Smooth Winterberry.

Frequent and locally rather plentiful. Wet places usually bordering small ponds. Western Cumberland and central and southern York Counties.

I. verticillata (L.) Gray. Winterberry, Black-alder.

Common, becoming less abundant north and northwestward; locally abundant. Low ground, wet woods, bogs, and other wet places. Throughout.

I. verticillata (L.) Gray forma **chrysocarpa** Robinson.

Local. With the species. Southeastern Washington, northern Kennebec, and northwestern Cumberland Counties.

I. verticillata (L.) Gray var. **cyclophylla** Robinson.

Local. With the species. Southeastern Hancock, east central Knox, and southern Lincoln Counties.

I. verticillata (L.) Gray var. **fastigiata** (Bicknell) Fernald.
See Rhod. 23:274. 1921.

Local. With the species. Southern Lincoln County.

I. verticillata (L.) Gray var. **padifolia** (Willd.) T. & G.

Local. With the species. Along the coast, southern Hancock and southern Lincoln Counties.

I. verticillata (L.) Gray var. **tenuifolia** (Torr.) Wats.

Infrequent. With the species. Southeastern Aroostook, southern Penobscot, eastern Washington, southeastern Hancock, northwestern Waldo, southwestern Knox, southern Lincoln, north central Kennebec, eastern Cumberland, and east central York Counties.

Nemopanthus. Mountain Holly.

N. mucronatus (L.) Trel. Mountain Holly.

Common and locally plentiful. Damp, cool woods and wet depressions in the mountains, frequently at high elevations; also in exposed rocky places along the coast where it occasionally attains a diameter of several inches. Throughout.

CELASTRACEAE (STAFF-TREE FAMILY)

Celastrus. Staff-tree, Shrubby Bittersweet.

C. scandens L. Shrubby or Climbing Bittersweet, Waxwork.

Rather frequent, becoming sparse northward. Thickets and along streams; cultivated and frequently persisting near old abandoned dwellings. Southern Piscataquis and south central Somerset Counties, and all portions of the State south of latitude 45, except Sagadahoc County.

ACERACEAE (MAPLE FAMILY)

Acer. Maple.

A. Negundo L. Box-elder, Ash-leaved Maple.

Indigenous to the Northeast but apparently introduced in the State. Commonly cultivated at an early date, especially in the French villages along the St. John River and its tributaries. Now generally escaped and locally established throughout in waste places and along rivers near habitation, but absent from the woods distant from urban centers.

A. pensylvanicum L. Striped or Goose-foot Maple, Moosewood.

Common in rich woods; occurring as a small understory tree. Throughout.

A. platanoides L. Norway Maple.
See Rehd. Man. ed. 2:568. 1940.

Native to Europe and western Asia. Introduced and commonly planted along the streets of the cities and villages; rarely escaping. Planted throughout, but less commonly northward.

A. platanoides L. var. **Schwedleri** Nichols. Schwedler Maple.
See Rehd. Man. ed. 2:569. 1940.

Occurrence and distribution as of the species but less commonly planted.

A. Pseudo-Platanus L. Sycamore Maple.
See Rehd. Man. ed. 2:574. 1940.

Native to Europe and western Asia. Introduced and occasionally planted as an ornamental in the portion of the State south of latitude 45.

A. rubrum L. Red Maple.

Common and widely distributed throughout the wooded areas; abundant in swamps; frequently planted along streets. Throughout.

A. rubrum L. var. **tomentosum** (Desf.) K. Koch.
See Rehd. Man. ed. 2:584. 1940.

Local. With the species. Station at Ocean Point, Southport, Lincoln County and Portland, Cumberland County.

A. rubrum L. var. **trilobum** K. Koch.
A. rubrum var. *tridens*. See Rehd. Man. ed. 2:584. 1940.

Occasional. With the species. Specimens from Knox, Lincoln, Androscoggin, Sagadahoc, and York Counties.

A. saccharinum L. Silver or White Maple.

Locally abundant in the southwestern part of the State, becoming infrequent northward. Restricted to the banks of streams and shores of lakes and ponds, but not in swamps. Occasionally planted as an ornamental but less commonly than the following variety. Throughout, except Washington, Hancock, Waldo, and Lincoln Counties.

A. saccharinum L. var. **laciniatum** (Carr.) Pax. Wier Maple.
See Rehd. Man. ed. 2:584. 1940.
Commonly planted horticultural form. Occasionally escaping and becoming established, especially along streams near habitation. Planted throughout, but more frequently in the southwestern part of the State.

A. saccharum Marsh. Sugar, Hard, or Rock Maple.
Abundant in rich woods; sparse elsewhere. Throughout. Commonly planted street tree.

A. spicatum Lam. Mountain Maple.
Common. Moist woods and cool north slopes especially in the vicinity of streams and ponds, usually associated with yellow birch or balsam fir. Throughout.

SAPINDACEAE (SOAPBERRY FAMILY)

Aesculus. Horse-chestnut, Buckeye.

A. glabra Willd. Ohio Buckeye.
Indigenous to the central states. Sparsely cultivated as an ornamental. Specimens planted and apparently hardy in Aroostook, Piscataquis, Somerset, Hancock, Waldo, Knox, and York Counties.

A. Hippocastanum L. Horse-chestnut.
Native to Europe. Introduced and commonly planted as an ornamental in the portion of the State south of latitude 45.

RHAMNACEAE (BUCKTHORN FAMILY)

Ceanothus. Red-root.

C. americanus L. New Jersey Tea.
Rare and local. Dry woodlands and gravelly banks and shores. Southern Penobscot and southeastern Oxford Counties.

Rhamnus. Buckthorn.

R. alnifolia L'Her. Alder-leaved Buckthorn.
Frequent and locally abundant, sparse along the coast. Mucky or peaty places or soils of calcareous origin. Swamps, alder thickets, and low, grassy swales and meadows near streams. Throughout, except Lincoln and Sagadahoc Counties.

R. cathartica L. Common Buckthorn.
Native to Europe. Commonly introduced and cultivated, principally for hedges. Often escaped and locally naturalized in thickets, fencerows, and along streams, especially near dwellings. Planted and found as an escape in southern Piscataquis County and practically throughout, south of latitude 45.

VITACEAE (VINE FAMILY)

Parthenocissus (*Psedera*). Virginia Creeper, Woodbine.

P. quinquefolia (L.) Planch. Woodbine, Virginia Creeper.
Psedera quinquefolia. See Rehd. Man. ed. 2:619. 1940.

Frequent, becoming sparse northward. Copses, thickets, and damp woods, especially near streams. Cultivated; often escaped and established along stone walls near dwellings. Throughout, south of latitude 46. Planted in northeastern and eastern Aroostook County and throughout its range southwestward. According to Fernald (Rhodora 41:430. 1939), essentially a southern species.

P. quinquefolia (L.) Planch. forma **hirsuta** (Donn) Fernald.
Psedera quinquefolia var. *hirsuta.* See Rhod. 41:429. 1939.

Local. With the species. Southern Penobscot, southern Lincoln, eastern Cumberland, and southern York Counties.

P. inserta (Kern.) K. Fritsch. Woodbine, Virginia Creeper.
Psedera vitacea. See Rehd. Man. ed. 2:619. 1940.

Frequent. Copses, moist woods, and alluvial thickets. Throughout, except the northern and central portions of Aroostook, Penobscot, Piscataquis, Somerset, Franklin, and Oxford Counties. Apparently absent from Washington and Sagadahoc Counties. According to Fernald (Rhodora 41:430. 1939), chiefly a northern and western species. Cultivated and often escaped to roadsides and stone walls near dwellings throughout its range.

P. inserta (Kern.) K. Fritsch forma **dubia** (Rehder) Fernald.
See Rehd. Man. ed. 2:619. 1940.

Local and infrequent. With the species. Southern Penobscot, west central Hancock, and east central Cumberland Counties.

Vitis. Grape.

V. labrusca L. Northern Fox Grape.

Frequent. Moist or dry thickets; also along roadsides and in waste places near habitation as an escape. Throughout, south of latitude 45, except central Franklin, north central Oxford, Washington, eastern Hancock, western Knox, and northern Lincoln Counties. Often cultivated for its fruits, and frequently escaping.

V. novae-angliae Fernald. New England or Pilgrim Grape.
See Rhod. 19:144-147. 1917.

Frequent and locally abundant. Principally in alluvial thickets and woods along the water courses. Also on steep banks and gravelly bars near the sea. Central and southern Penobscot, southern Piscataquis, central and southern Somerset, central and southern Franklin, southern Oxford, west central Hancock, northeastern Waldo, east central Knox, northern Lincoln, northeastern Kennebec, central Sagadahoc, central and southern Cumberland, and southeastern York Counties.

V. riparia Michx. River-bank or Frost Grape.
V. vulpina. See Rhod. 41:431-434. 1939.

Frequent and locally abundant, especially along the rivers. Stream banks, thickets, borders of woods and roadsides. Northeastern Aroostook, Penobscot, southern Piscataquis, southern Somerset, southern Franklin, west central Oxford, Waldo, northwestern Knox, southern Lincoln, Kennebec, Androscoggin, east central Sagadahoc, southern Cumberland, and northeastern York Counties.

TILIACEAE (LINDEN FAMILY)

Tilia. Basswood. Linden.

T. americana L. American Basswood, American Linden.

Common, becoming infrequent northward. Rich woods. Throughout, except northern Aroostook, northern Penobscot, northern and central Piscataquis, northern Somerset, northern Franklin, and northern Oxford Counties. Frequently used as an ornamental. The northern species often referred to as *T. glabra* Vent. (For validity of name see Leon Croizat in Torreya 37:55-57. 1937.)

T. cordata Mill. Small-leaved European Linden.
See Rehd. Man. ed. 2:624. 1940.

Native to Europe. Introduced as an ornamental and shade tree. Occasionally planted in western Kennebec, west central Knox, southwestern Sagadahoc, eastern Cumberland, and central York Counties.

× **T. europaea** L. Common Linden.
See Rehd. Man. ed. 2:623. 1940.

Native to Europe. Introduced as an ornamental and shade tree. Occasionally planted in southern Penobscot, southwestern Piscataquis, central Waldo, southern Cumberland, and south central York Counties.

CISTACEAE (ROCKROSE FAMILY)

Helianthemum. Rockrose.

H. Bicknellii Fernald. Hoary Frostweed.
See Rhod. 21:36-37. 1919.

Infrequent, becoming sparse northward. Sandy or gravelly soil in open places. Southern Penobscot, southern Somerset, southern Franklin, eastern Oxford, northeastern Kennebec, northern Androscoggin, western Sagadahoc, Cumberland, and York Counties.

H. canadense Michx. Frostweed.

Infrequent, becoming sparse northward. Sandy and gravelly soil in open places. Southern Penobscot, southern Franklin, eastern Oxford, central Knox, western Sagadahoc, eastern and southern Cumberland, and York Counties.

Hudsonia.

H. ericoides L. Heath-like Hudsonia.

Infrequent and local. Dry sandy soil, rock crevices and exposed places on the mountains along the coast. Southern Oxford, and along the coast in southern Washington, eastern Hancock, eastern Knox, central and eastern Sagadahoc, eastern Cumberland, and east central York Counties.

H. tomentosa Nutt. Woolly Hudsonia.

Infrequent and locally abundant. Sandy shores and beaches. Southern Oxford, central and southern Sagadahoc, extreme northwestern and southeastern Cumberland, and eastern York Counties.

H. tomentosa Nutt. var. **intermedia** Peck.

Local. With the species. Southwestern Oxford and northwestern Cumberland Counties.

THYMELAEACEAE (Mezereum Family)

Daphne. Mezereum.

D. Mezereum L. Mezereum.

Native to Europe. Occasionally escaped from cultivation and locally established near dwellings. Planted throughout, south of latitude 45, except Washington, Hancock, and Sagadahoc Counties.

Dirca. Leatherwood.

D. palustris L. Leatherwood, Wicopy.

Frequent and locally rather abundant, especially on the calcareous soils. East central and southern Aroostook, Penobscot, southern Piscataquis, southern Somerset, southern Franklin, eastern Oxford, eastern Washington, northwestern Knox, Kennebec, Androscoggin, central Sagadahoc, Cumberland, and York Counties.

ELAEAGNACEAE (Oleaster Family)

Elaeagnus. Oleaster.

E. angustifolia L. Russian Olive, Oleaster.
See Rehd. Man. ed. 2:664. 1940.

Native to Europe and Asia. Introduced as an ornamental and rarely escaped. Planted in southern Penobscot, southwestern Piscataquis, southern Somerset, north central Waldo, west central Knox, central Lincoln, central and northern Kennebec, southeastern Cumberland, and southeastern York Counties.

Shepherdia.

S. canadensis (L.) Nutt. Russet Buffalo-berry.

Local. Rocky cliffs. Madison (Fernald, August 1, 1892, and Hyland No. 943, July 9, 1938), Somerset County.

Arthur H. Norton recently called attention to a statement by Ora W. Knight (The Birds of Maine, p. 85, Charles H. Glass & Co., Bangor, Maine, 1908) on the possible occurrence of the Buffalo-berry near Bucksport, Hancock County. Knight, in discussing the Red-legged Black Duck, writes: "Two individuals [ducks] which were killed in winter on the Penobscot River near Bucksport had been feeding on a peculiar red berry about eight millimeters in diameter, containing much seed and little pulp. The birds were literally crammed with these fruits which were unknown to me and which have never been identified but which were most certainly not any common fruit in Maine or I would have known them...... Since writing this I have been able to positively identify the fruit as that of *Lepargyraea Canadensis* Nutt. [*Shepherdia canadensis* (L.) Nutt.], a northern shrub not known from this particular region."

LYTHRACEAE (Loosestrife Family)

Decodon. Swamp Loosestrife.

D. verticillatus (L.) Ell. Water-willow.

Frequent and locally plentiful. Swampy places and along the shores of streams and ponds, often growing partly submersed throughout the season. Southern Penobscot, southern Franklin, east central and southern Oxford, southeastern Hancock, east central Waldo, west central Knox, southern Lincoln, Kennebec, southwestern Androscoggin, western Cumberland, and York Counties.

D. verticillatus (L.) Ell. var. **laevigatus** T. & G.
See Torr. & Gray Fl. 1:483. 1840.
  Infrequent. With the species. Southern Penobscot, southern Franklin, southeastern Hancock, eastern and west central Knox, southern Lincoln, and west central Kennebec Counties.

D. verticillatus (L.) Ell. var. **pubescens** T. & G.
See Torr. & Gray Fl. 1:483. 1840.
  Local. With the species. Chiefly near the coast. (Reported in Rhodora 19:154-155. 1917.)

ARALIACEAE (GINSENG FAMILY)

Aralia.

A. hispida Vent. Prickly or Bristly Sarsaparilla.
  Common but becoming sparse northward. Locally abundant in clearings, especially on old burns; also in rich woods, in rock crevices, and on gravelly banks along highways. Throughout.

A. nudicaulis L. Wild Sarsaparilla.
  Common. Moist and dry woodlands and on gravelly banks; often persisting under pine. Throughout.

A. spinosa L. Hercules' Club, Angelica-tree, Devil's Walking-stick.
  Indigenous to the central and southeastern states. Introduced and occasionally planted as an ornamental but rarely, if ever, persisting as an escape. Southern Penobscot, southern Hancock, northern Knox, southeastern Cumberland, and northeastern York Counties.

CORNACEAE (DOGWOOD FAMILY)

Cornus. Dogwood.

C. alternifolia L. f. Pagoda, Blue, or Alternate-leaved Dogwood.
  Common but nowhere abundant. Copses and dry rocky banks, borders of fields and woods. Throughout.

C. Amomum Mill. Silky Dogwood or Cornel.
  Common, becoming less frequent along the coast northeastward. Principally restricted to wet places along the water courses. Southwestern Aroostook and throughout, south of latitude 45.5, except the eastern portions of Washington, Hancock, and Knox Counties.
  A segregate (*C. Purpusi* Koehne) is recognized by some botanists but is not here considered as distinct. (For discussion see Rhodora 6:55-56. 1904.)

C. florida L. Flowering Dogwood.
  Local. Dry rocky woods. Mt. Agamenticus, York, York County. Occasionally planted as an ornamental in southern Penobscot, southern Hancock, west central Lincoln, and eastern Cumberland Counties.
  Apparently the specimens collected by Dr. Anne E. Perkins in 1936 and 1937 (Rhodora 40:463. 1938) are the first in the State and represent the sole Maine stations as now known. Reported by Blakeslee and Jarvis (Trees in Winter, p. 418. 1913. The Macmillan Co.) from Fayette Ridge, Winthrop, Kennebec County; however, no specimens are at present available.

C. racemosa Lam. Panicled or Gray Dogwood.
C. paniculata. See Rehd. Man. ed. 2:685-686. 1940.

Infrequent to frequent. Thickets and banks of streams. Southern Penobscot, central and southern Somerset, southeastern Hancock, western Knox, Kennebec, southeastern Cumberland, and northeastern and southern York Counties.

C. rugosa Lam. Round-leaved Dogwood or Cornel.
C. circinata. See Rehd. Man. ed. 2:684. 1940.

Common, becoming less frequent northward, especially in the mountainous portions. Copses, borders of rich woods; also in sandy soil and on rocks. Throughout.

C. rugosa × stolonifera.

Local. Represented in the State by a single collection (Fernald No. 305) as reported by Rehder (Rhodora 12:123. 1910). Gravelly shores, valley of the Piscataquis River, Dover, Piscataquis County.

C. stolonifera Michx. Red-osier Dogwood.

Common, infrequent along the coast and in the deep woods northward. Along the water courses and wet places generally. Throughout. (See Bull. Torr. Bot. Club 69:583-588. 1942, for a discussion of name.)

Nyssa. Tupelo, Pepperidge, Sour-gum.

N. sylvatica Marsh. Black Tupelo, Black-gum.

Frequent, becoming local northward. Deep, rich, moist or nearly dry soil bordering woods and streams and occasionally in swamps. Northern and western Kennebec, northern and southwestern Androscoggin, Sagadahoc, Cumberland, and York Counties. Specimens from Kennebec County were reported as early as 1867 (Bessey, American Naturalist 15:134. 1881).

ERICACEAE (Heath Family)

Andromeda. Bog Rosemary.

A. glaucophylla Link. Bog Rosemary.

Locally abundant. Bogs and wet shores. Throughout, except Waldo and Sagadahoc Counties.

Arctostaphylos. Bearberry.

A. alpina (L.) Spreng. Alpine Bearberry.

Local. High mountain summits. Mt. Katahdin, Piscataquis County.

A. Uva-ursi (L.) Spreng. var. **coactilis** Fernald & MacBride. Bearberry.
See Rhod. 16:212. 1914.

Infrequent, but locally abundant, becoming sparse northward. Bare hills and rocks, often at high elevations; also in sandy barrens under pine. T. 11, R. 8, Aroostook County, Kineo, Piscataquis County, Flagstaff, Somerset County, and throughout, south of latitude 45, except northern and central Oxford, and Waldo Counties.

Calluna. Heather.

C. vulgaris (L.) Hull. Heather.

Native to Europe. Introduced; cultivated and occasionally escaped and well established in open places near habitation. Central Franklin, eastern Oxford, and along the coast in Cumberland County.

Cassiope.

C. hypnoides (L.) D. Don.
Local. Protected upper slopes near the higher mountain summits. Mt. Katahdin, Piscataquis County.

Chamaedaphne. Leather-leaf.

C. calyculata (L.) Moench. Leather-leaf.
Common and locally abundant, becoming less frequent northward. Abundant in bogs and along wet shores; also in wet places at high elevations near mountain summits. Throughout.

Chimaphila. Pipsissewa.

C. umbellata (L.) Nutt. var. **cisatlantica** Blake. Pipsissewa, Prince's-pine.
See Rhod. 19:241-242. 1917.
Common, becoming somewhat less frequent northward. Dry woods; often on small knolls under pine and hemlock. Throughout.

Chiogenes. Creeping Snowberry.

C. hispidula (L.) T. & G. Creeping Snowberry, Moxie-plum, Capillaire.
Common and locally abundant. Peat bogs and damp, mossy woods; also in wet places in the mountains at high elevations. Throughout.

Clethra. White-alder.

C. alnifolia L. White-alder, Sweet Pepperbush.
Local. Wet copses, moist woods, and borders of ponds; often in large patches. Northwestern Cumberland and central and southern York Counties. Planted as an ornamental in southern Penobscot, southern Hancock, east central Waldo, north central Knox, and eastern Cumberland Counties.

Epigaea. Trailing Arbutus.

E. repens L. var. **glabrifolia** Fernald. Trailing Arbutus, Mayflower.
See Rhod. 41:444-446. 1939.
Common. Sandy or rocky woods, especially on knolls under pines. Throughout.

E. repens L. var. **glabrifolia** Fernald forma **plena** Rehder.
See Rehd. Man. ed. 2:737. 1940.
Two stations reported for the State (Rhodora 38:408. 1936) without specific localities.

Gaultheria. Aromatic Wintergreen.

G. procumbens L. Checkerberry, Teaberry.
Common and locally plentiful. Rather dry woods, often on humus-covered rocks and in clearings and gravelly or rocky places in the mountains; also in seaside meadows and peaty places on ledges along the coast. Throughout.

Gaylussacia. Huckleberry.

G. baccata (Wang.) C. Koch. Black Huckleberry.
Common and locally abundant. Rocky woods, swamps, and bogs. Throughout, south of latitude 46.

G. baccata (Wang.) C. Koch forma **glaucocarpa** (Robinson) Mackenzie.
Blue Huckleberry.
Local. With the species. West central Oxford, east central Hancock, and southern York Counties.

G. baccata (Wang.) C. Koch forma **leucocarpa** (Porter) Fernald.
White Huckleberry.
Local. With the species. Eastern Cumberland County.

G. dumosa (Andr.) T. & G. var. **Bigeloviana** Fernald. Dwarf Huckleberry.
See Rehd. Man. ed. 2:747. 1940.
Infrequent and local. Sandy swamps mostly along the coast. Southern Penobscot, southern Washington, southeastern Hancock, northern Lincoln, central Kennebec, and central and southern York Counties.

Kalmia. Laurel.

K. angustifolia L. Lamb-kill, Sheep-laurel.
Common and abundant, becoming less frequent northward. Hillsides, old pastures, and bogs; also on humus-covered rocks in the mountains at high elevations. Throughout.

K. angustifolia L. forma **candida** Fernald.
See Rehd. Man. ed. 2:727. 1940.
Occasional with the species but probably seldom considered as distinct. Ledgy pasture. North Waldoboro, Waldoboro (Hyland No. 69), Lincoln County.

K. latifolia L. Mountain-laurel.
Local and rather widely distributed. Rocky woods and low ground. Southern Aroostook (report), southern Penobscot, southwestern Oxford, southern Washington, southeastern Hancock, southern Sagadahoc, Cumberland, and York Counties. Planted as an ornamental in northeastern Aroostook, southern Penobscot, southwestern Piscataquis, central and southern Oxford, southern Hancock, southern Waldo, west central Knox, west central Lincoln, and eastern Cumberland Counties.

Note: An act for the protection of *Rhododendron maximum* Linnaeus and *Kalmia latifolia* Linnaeus:
BE IT ENACTED BY THE PEOPLE OF THE STATE OF MAINE, as follows: **Certain plants protected.** Whoever without the consent of the owner of the land whereon the same may be growing injures, destroys, digs up or removes any *Rhododendron maximum* Linnaeus or *Kalmia latifolia* Linnaeus, or any part or parts of the plants of either of said species growing upon the land of another shall be guilty of a misdemeanor and shall be punished by a fine of not more than $100, and in addition thereto shall be liable to the owner of the land upon which the same was growing in an action of trespass in treble damages.
Approved March 25, 1937.
Laws of Maine 1937, Kennebec Journal, Augusta, Maine, 1937. Public Laws of the State of Maine Eighty-Eighth Legislature 1937, Chapter 71, page 90.

K. polifolia Wangenh. Pale-laurel.

Locally common, becoming somewhat less frequent northward. Restricted principally to cold bogs but found also in wet depressions in the mountains. Throughout, except Lincoln and Sagadahoc Counties.

Ledum. Labrador Tea.

L. groenlandicum Oed. Labrador Tea.

Common and locally abundant. Bogs, damp thickets, hills, seepy hillsides, and cold mountain slopes. Throughout.

Loiseleuria. Alpine-azalea.

L. procumbens (L.) Desv. Alpine-azalea.

Local. Sheltered upper slopes near the higher mountain summits. Mt. Katahdin, Piscataquis County.

Lyonia.

L. ligustrina (L.) DC. Male-berry.

Common and locally abundant, becoming sporadic northward. Moist thickets and borders of swamps and bogs. Southern Penobscot and west central Piscataquis Counties, and all sections of the State south of latitude 45, except Washington and Hancock Counties.

L. ligustrina (L.) DC. var. **foliosiflora** (Michx.) Fernald.

Local and apparently rare. With the species. Southwestern Kennebec and central Androscoggin Counties.

Phyllodoce.

P. coerulea (L.) Bab. Mountain Heath.

Local. Protected upper slopes near the higher mountain summits. Mt. Katahdin, Piscataquis County and Grafton, Oxford County.

Rhododendron.

R. canadense (L.) BSP. Rhodora.

Common and locally abundant, becoming somewhat infrequent northward. Swamps, bogs, and damp slopes. Throughout.

R. canadense (L.) BSP. forma **albiflorum** (Rand & Redf.) Rehd.

See Fl. Mt. Desert Island, Maine, p. 127. 1894.

Occasional with the species. Reported by Rand and Redfield (l.c.) from Southwest Harbor, and by nurserymen at Hancock and Ellsworth, Hancock County.

R. lapponicum (L.) Wahlenb. Lapland Rhododendron or Rose-bay.

Local. Upper slopes near the higher mountain summits. Mt. Katahdin, Piscataquis County.

R. maximum L. Rose-bay Rhododendron, Great-laurel.

Rare and local. Damp, deep woods, mostly near ponds or lakes. Apparently not on limestone soils. Southwestern Somerset, southern Cumberland, and southwestern York Counties. Occasionally planted in northeastern Aroostook, southern Penobscot, and Washington Counties and southwestward, principally in the counties bordering the coast.

Protected by law. (See note on legislative action under Mountain-laurel.)

R. viscosum (L.) Torr. Clammy Azalea.

Rare and local. Swamps, mostly near the coast. Northeastern Cumberland and and southeastern York Counties. Planted in southwestern Piscataquis County and occasionally in the area south of latitude 44.5.

R. viscosum (L.) Torr. var. **glaucum** (Michx.) Gray.

Local. With the species. Wells (Parlin No. 800), York County.

Vaccinium. Bilberry, Blueberry, Cranberry.

The treatment of this genus follows that of Rehder's Manual of Cultivated Trees and Shrubs, ed. 2, 1940, supplemented by joint field work with W. H. Camp (1941) and correspondence (1942).

V. angustifolium Ait. Low Sweet, Early Sweet, or Lowbush Blueberry.

V. pensylvanicum var. *angustifolium.* See Rehd. Man. ed. 2:751. 1940.

Abundant. Barrens, dry hills, and other open or sparsely shaded places; also in heaths and in the mountains at high elevations. Throughout. The most widespread and commercially the most important blueberry of the State. According to Camp the diploid phase is more common than the tetraploid. They are difficult to separate.

The albino-fruited form [*V. angustifolium* forma *leucocarpum* (Deane) Rehd., 1.c.] has been collected locally at Jordan Mountain, Mt. Desert Island (E. L. Rand, Sept. 2, 1892), Hancock County.

V. angustifolium × **caesariense.**

Local. With the parents. Howland (Hyland No. 1019), Penobscot County.

V. angustifolium × **corymbosum.**

Frequent, occurring throughout the range of the parents. These are the familiar "half-high" plants. In areas where the tetraploid phase of *angustifolium* and *corymbosum* come together, hybrids, hybrid segregates, and back-crosses are to be found; these more or less intermediate plants have been called *V. atlanticum* Bicknell (Bull. Torr. Bot. Club 41:422. 1914).

V. angustifolium Ait. var. **myrtilloides** (Michx.) House.

V. pensylvanicum var. *myrtilloides.* See Rhod. 10:147-148. 1908.

Infrequent. With the species. Northwestern and central Aroostook, southern Penobscot, east central Piscataquis (fide W. H. Camp, 1942), west central Somerset (report), Franklin, eastern Washington, south central Knox, and southwestern Androscoggin Counties.

Apparently a series of hybrids between *angustifolium* and *canadense*, together with their segregates.

V. angustifolium Ait. var. **nigrum** (Wood) Dole. Low Black-fruited Shiny-leaved Blueberry.

V. pensylvanicum var. *nigrum.* See Dole, Fl. Vermont, ed. 3 rev., p. 210. 1937; Rehd. Man. ed. 2:751. 1940.

Common and locally abundant. With the species or by itself. Throughout, except Aroostook County.

Apparently often confused with *V. Brittonii* Porter and other entities.

V. atrococcum (Gray) Heller. Black Highbush or Downy Swamp Blueberry.

Locally abundant, becoming infrequent northward. Swamps, bogs, low woods, and rocky pastures. Southern Penobscot, west central Somerset, Hancock, Knox, southern Lincoln, southwestern Androscoggin, Cumberland, and York Counties.

V. Brittonii Porter. Low Black-fruited Pale-leaved Blueberry.
See Bull. Torr. Bot. Club 41:420. 1914.

Frequent and probably widespread, often occurring with *V. angustifolium*. Complete data on range and distribution are at present lacking principally because of confusion with other entities. Apparently throughout. Specimens recently collected at Swanville (G. D. Chamberlain No. 1905), Waldo County, have been identified by Dr. Camp as typical.

V. caesariense Mackenzie.

See Torreya 10:230. 1910.

Frequent in the southern portion of the State, becoming infrequent northward. Complete data on range and distribution are at present lacking. Specimens have been collected as far north as Howland (Hyland No. 1020), Penobscot County.

V. caespitosum Michx. Dwarf Bilberry.

Locally frequent to rather common in the northern and west central parts of the State. Occurring as a disjunct along the coast in southern York County. (See Rhodora 2:187-190. 1900.) Open gravelly or rocky hills and woods, sandy knolls, ledgy banks of streams and exposed mountain summits. Aroostook, Penobscot, Piscataquis, Somerset, Franklin, Oxford, and southern York Counties.

V. canadense Kalm. Canada, Sour-top, or Velvet-leaf Blueberry.

Common and locally abundant. Dry plains, barrens, swamps, bogs, and moist or rocky woods; also in the mountains. Throughout.

At the suggestion of Dr. Camp (personal correspondence Aug. 6, 1942) the following statement is added to tentatively take care of certain specimens as yet unnamed: "In the Katahdin region and in other areas are taller-than-typical plants which are suspected polyploids."

V. canadense Kalm forma **chiococcum** Deane.

Local. With the species. Southern Penobscot and southeastern Hancock Counties.

V. corymbosum L. Highbush or Swamp Blueberry.

Common and locally abundant, becoming infrequent northward. Swamps, bogs, low woods, and rocky pastures. Throughout, south of latitude 46, except northern Penobscot, central Piscataquis, central Somerset, northern and central Franklin, northern and central Oxford, and eastern Washington Counties.

V. corymbosum L. forma **albiflorum** (Hook.) Camp.

V. corymbosum var. *amoenum*. See Am. Midland Nat. 23:177. 1940.

Frequent. With the species. Northeastern Washington County and throughout, south of latitude 45, except southern Somerset, northern and central Oxford, Waldo, and Sagadahoc Counties.

V. corymbosum L. forma **glabrum** (Gray) Camp.

V. corymbosum var. *pallidum*. See Am. Midland Nat. 23:177. 1940.

Infrequent. With the species. Southern Penobscot, east central Oxford, southeastern Hancock, northeastern Cumberland, and southern York Counties.

V. macrocarpon Ait. Large or American Cranberry.

Rather common and locally abundant, becoming less frequent northward. Open bogs, swamps, wet shores, and low meadows; also on humus-covered rocks near the coast. Throughout, except western Aroostook, northern Penobscot, northern Piscataquis, northern Somerset, northern Franklin, and northern Oxford Counties.

V. Oxycoccus L. Small Cranberry.

Frequent to common, becoming locally abundant along the coast. In sphagnum and wet humus. Throughout, except western and west central Aroostook, northern Penobscot, northern Piscataquis, central Somerset, northern Oxford, northern Washington, and York Counties.

V. Torreyanum Camp. Dryland or Late Low Blueberry.
V. vacillans in part. See Am. Midland Nat. 23:177. 1940.

Infrequent, becoming local northward. Dry places, mostly in sandy soil or rocky woods. Southeastern Penobscot, central and southern Oxford, southern Kennebec, Cumberland, and York Counties.

Regarding the nomenclature of this species Dr. Camp (personal correspondence Aug. 6, 1942) states: "It is my opinion your material in Maine is a form of *Tôrreyanum* rather than the much coarser *pallidum* of the middle and southern Appalachian." (See also Fernald in Rhodora 45:456-457. 1943 for a discussion of name.)

V. uliginosum L. var. **alpinum** Bigel. Bog or Mountain Bilberry.
V. uliginosum. See Rhod. 25:23-24. 1923.

Frequent and locally abundant in the mountainous regions and occasionally at lower elevations. (See Rhodora 2:187-190. 1900.) Chiefly on the ledgy slopes above the timber line on the higher mountains but occasionally on dry hillsides and ledgy shores. Northern Aroostook, east central and southern Penobscot, central Somerset, Franklin, and central Oxford Counties.

A variety (*V. uliginosum* L. var. *pubescens* Lange) is sometimes recognized as distinct but within our borders, according to Fernald (Rhodora 25:23-24. 1923), it does not seem to be satisfactorily separable. Specimens (so labeled on herbarium sheets) have frequently been collected with the species in Piscataquis and Oxford Counties.

V. Vitis-idaea L. var. **minus** Lodd. Mountain or Rock Cranberry.

Locally abundant along the coast and on the mountains inland. Dry or rocky banks, seaside meadows, mountain slopes, and alpine summits; rarely in wet moss. Northern Penobscot, east central and southwestern Piscataquis, central and southeastern Somerset, Franklin, and central Oxford Counties and principally near the coast in Washington, Hancock, Waldo, Knox, Sagadahoc, Androscoggin (southwestern portion), Cumberland, and York Counties.

DIAPENSIACEAE (Diapensia Family)

Diapensia.

D. lapponica L.

Local. Tablelands and high mountain summits. Mt. Katahdin, Piscataquis County, Mt. Saddleback and Mt. Abraham, Franklin County, and Grafton, Oxford County.

OLEACEAE (Olive Family)

Fraxinus. Ash.

F. americana L. White Ash.

Common forest tree. Plentiful in rich, moist woods, associated with other hardwoods. Throughout.

F. americana L. forma **iodocarpa** Fernald.
See Rhod. 14:192. 1912.

Occasional. With the species. Southern Hancock and southern Kennebec Counties.

F. nigra Marsh. Black Ash.

Common and locally abundant. Occasionally along wet shores but mostly restricted to swamps where it is associated with red maple. Throughout.

F. pennsylvanica Marsh. Red Ash.

Frequent and locally abundant. Low ground, especially inhabiting the banks along the water courses. Often associated with silver maple. Northern Aroostook, central and southern Penobscot, southern Somerset, southern Oxford, eastern Washington (report), southern Waldo, northwestern Knox, central Lincoln, Kennebec, Androscoggin, eastern Sagadahoc, and southern and northwestern Cumberland Counties.

F. pennsylvanica Marsh. var. **Austinii** Fernald.

See Rhod. 40:452-453. 1938.

Infrequent and rather widely distributed. With the species but generally more northern. Northern Aroostook, eastern and central Penobscot, southern Somerset, southern Franklin, northeastern Washington, southern Hancock, western and southern Knox, northern Kennebec, southern Androscoggin, western Sagadahoc, eastern Cumberland, and northeastern York Counties.

F. pennsylvanica Marsh. var. **lanceolata** (Borkh.) Sarg. Green Ash.

Infrequent and local. Essentially along the water courses. Southern Penobscot, southern Franklin, northern Kennebec, and northeastern Androscoggin Counties.

Syringa. Lilac.

S. vulgaris L. Common Lilac.

Native to Europe. Commonly cultivated as an ornamental. Occasionally established in waste places near dwellings and old house sites; common in city dumps. Escaped and established throughout, except in the wooded portions of western Aroostook, northern Penobscot, northern Piscataquis, northern and central Somerset, and northern Oxford Counties.

POLEMONIACEAE (POLEMONIUM FAMILY)

Phlox.

P. subulata L. Ground- or Moss-pink.

Indigenous to New York and southwestward. Introduced as an ornamental and sparingly naturalized, especially in the vicinity of cemeteries. Planted in southern Penobscot, southern Somerset, central Kennebec, central Sagadahoc, east central Cumberland, and southern York Counties.

LABIATAE (MINT FAMILY)

Thymus. Thyme.

T. Serpyllum L. Creeping Thyme.

Native to Europe, western Asia and northern Africa. Occasionally introduced as an ornamental, frequently spreading to old fields and waste places near dwellings. Established in southwestern Penobscot, southern Franklin, southern Hancock, north central Kennebec, northern Androscoggin, and Cumberland Counties.

SOLANACEAE (NIGHTSHADE FAMILY)

Lycium. Matrimony-vine.

L. halimifolium Mill. Common Matrimony-vine.

Native to Europe. Introduced as an ornamental and occasionlly escaped to waste grounds about dwellings in Mt. Desert, Hancock County and South Berwick,

York County. Frequently planted in practically all sections of the State south of latitude 45.

Solanum. Nightshade.

S. Dulcamara L. Climbing or Bitter Nightshade, European Bittersweet.

Native to Europe and Asia. Introduced and now widely naturalized. Moist banks and waste places near dwellings. Occasionally found bordering ponds and streams, often growing in shallow water. Southern Piscataquis County and throughout, south of latitude 45, except western Oxford, western Washington, northern Hancock, and Lincoln Counties.

S. Dulcamara L. var. **villosissimum** Desv.

See Rehd. Man. ed. 2:813. 1940.

Native to Europe and Asia. Introduced, occasionally escaping and becoming naturalized. Infrequent. With the species. Southeastern Franklin, northeastern Cumberland, and southern York Counties.

BIGNONIACEAE (BIGNONIA FAMILY)

Catalpa. Catalpa, Indian-bean.

C. bignonioides Walt. Southern or Common Catalpa, Indian-bean, Cigar-tree.

Indigenous to the Gulf coast but naturalized as far north as New York. Introduced and occasionally planted as an ornamental in the eastern part of the State. Southern Penobscot, southern Washington, southern Sagadahoc, Cumberland, and central and eastern York Counties.

C. speciosa Warder. Northern Catalpa.

Indigenous to the south central states. Introduced and occasionally planted as an ornamental. Southern Penobscot, southern Piscataquis, southern Somerset, eastern Washington, Knox, northern Lincoln, Kennebec, Androscoggin, Sagadahoc, Cumberland, and York Counties.

RUBIACEAE (MADDER FAMILY)

Cephalanthus. Buttonbush.

C. occidentalis L. Common Buttonbush.

Locally abundant; sparse in the mountains northwestward and along the coast northeastward. Swamps and borders of lakes and streams; the lower portions of the plant often submersed throughout the season. Southeastern Aroostook, southern Penobscot, east central Washington, and northern Hancock Counties, and throughout the State south of latitude 45.

Mitchella. Partridge-berry.

M. repens L. Partridge-berry.

Common and locally rather abundant. Dry woods, especially under pines. Throughout.

CAPRIFOLIACEAE (HONEYSUCKLE FAMILY)

Diervilla. Bush-honeysuckle.

D. Lonicera Mill. Bush-honeysuckle.

Common. Dry woods, borders of fields, roadsides, and rocky places generally. Throughout.

Linnaea. Twin-flower.

L. borealis L. var. **americana** (Forbes) Rehd. Twin-flower.

Common and locally abundant. Moist mossy woods and cold bogs; also in the mountains. Throughout.

Lonicera. Honeysuckle.

L. canadensis Marsh. American Fly Honeysuckle.

Common. Chiefly in moist or rather dry woods, often tolerating dense shade. Throughout.

L. dioica L. Climbing Honeysuckle.

Local and rather rare. Rocky or sandy grounds. Central Kennebec, southern Sagadahoc, eastern Cumberland, and west central York Counties.

L. Morrowi Gray. Morrow Honeysuckle.

Native to Japan. Introduced and occasionally escaped in southern Penobscot and eastern and central Knox Counties. Frequently planted as an ornamental in southeastern Aroostook County and in the urban centers throughout, south of latitude 45.

L. oblongifolia (Goldie) Hook. Swamp Fly Honeysuckle.

Locally plentiful. Open places in cool arbor-vitae and tamarack swamps underlaid with limestone; occasionally along peaty or boggy banks of streams elsewhere. Aroostook, northern and east central Penobscot, and west central Piscataquis Counties.

L. sempervirens L. Trumpet Honeysuckle.

Rare and local. Copses. Northeastern Cumberland and York Counties. Often cultivated as an ornamental and planted in southern Penobscot, southwestern Piscataquis, west central Franklin, eastern Washington, southern Hancock, southern Waldo, west central Knox, central Kennebec, and eastern York Counties.

L. tatarica L. Tatarian Honeysuckle.

Native to Asia. Introduced and commonly planted, frequently escaping and becoming established in places distant from habitation. Roadsides, rocky shores, and sparsely wooded banks. Southern Penobscot, southern Piscataquis, east central Washington, north central Knox, northeastern Kennebec, west central Sagadahoc, eastern Cumberland, and east central York Counties.

L. villosa (Michx.) R. & S. Mountain Fly Honeysuckle.

L. caerulea var. *villosa* in part. See Rhod. 27:5-6. 1925.

Frequent and locally plentiful. Rocky pastures, low woods, and bogs. Southern Penobscot, Hancock, southwestern Kennebec, and east central York Counties.

L. villosa (Michx.) R. & S. var. **calvescens** (Fernald & Wieg.) Fernald.
See Rhod. 27:8-9. 1925.

Widely distributed. Sphagnum bogs, grassy swales, and shelves at high elevations in the mountains. East central Piscataquis, northwestern Oxford, southeastern Washington, eastern Knox, and southwestern Androscoggin Counties.

L. villosa (Michx.) R. & S. var. **Solonis** (Eaton) Fernald.
See Rhod. 27:6-8. 1925.

Frequent and widely distributed. Sphagnum bogs, damp woods, sand plains, dry rocks, and in the mountains at rather high elevations. Aroostook, Penobscot, east central Piscataquis, northern Somerset (report), southern Franklin, east central Oxford, Washington, southeastern Hancock, Kennebec, northern Androscoggin, and northeastern Cumberland Counties.

L. villosa (Michx.) R. & S. var. **tonsa** Fernald.
See Rhod. 27:9-10. 1925.

Infrequent and rather widely distributed. Moist thickets, grassy swales, wet shores, and rocky places. Northeastern Aroostook, east central Penobscot, west central Franklin, Washington, Hancock, and southern Knox Counties.

L. Xylosteum L. European Fly Honeysuckle.

Native to Europe. Introduced as an ornamental and occasionally escaped near dwellings in Skowhegan, Somerset County and Portland, Cumberland County.

Sambucus. Elder.

S. canadensis L. Common, American or Black-berried Elder.

Common, becoming infrequent northwestward. Rich soil. Woods, thickets, fencerows, and open places generally; often near streams. Throughout.

S. pubens Michx. Red-berried Elder.
S. racemosa. See Rhod. 35:310. 1933.

Common. Open or rocky woods, often near streams. Throughout.

S. pubens Michx. forma **calva** Fernald.
See Rhod. 35:310. 1933.

Sporadic. With the species. North central Franklin, southern Hancock, and southern Lincoln Counties.

Symphoricarpos. Snowberry.

S. albus (L.) Blake var. **laevigatus** (Fernald) Blake. Snowberry.
S. racemosus var. *laevigatus.* See Rhod. 16:117-119. 1914.

Indigenous to New England and adjacent Canada, but apparently occurring in the State only under cultivation or as an escape. Roadsides and waste places near dwellings. Commonly planted and frequently escaped throughout, except in the wooded portions of northern and western Aroostook, northern Penobscot, northern Piscataquis, northern and central Somerset, and northern Oxford Counties.

Viburnum. Arrow-wood.

V. acerifolium L. Maple-leaved Arrow-wood, Dockmackie.

Frequent to rather common, becoming infrequent northward. Chiefly in dry or rocky woods, mostly under conifers. Central Somerset County and throughout, south of latitude 45.

V. alnifolium Marsh. Hobble-bush, Witch Hobble, Moosewood.

Frequent to common. Moist woods, often in medium or dense shade. Throughout.

V. cassinoides L. Withe-rod, Wild Raisin.

Common and locally plentiful. Swamps, sparse or low woods, and open places on wet to dry soil; also at rather high elevations in the mountains. Throughout. Occasionally cultivated.

V. dentatum L. var. **lucidum** Ait. Arrow-wood.

V. dentatum. See Rhod. 42:1-6. 1940.

Common and locally plentiful. Thickets, wet or dry woods, borders of fields and other open places; frequent on banks along the water courses. Throughout, south of latitude 46. Often cultivated.

V. edule (Michx.) Raf. Squashberry, Mooseberry, Pimbina.

V. pauciflorum. See Rhod. 43:481-483. 1941.

Rather rare and local. Cold woods, usually near streams; also at high elevations in the mountains. Northern and western Aroostook, northern Penobscot, east central and southwestern Piscataquis, northern and west central Somerset, Franklin, and central Oxford Counties.

V. Lentago L. Nanny-berry, Sweet Viburnum, Wild Raisin.

Frequent and locally plentiful, becoming infrequent northward and along the coast northeastward. Principally along the water courses but occasionally in low woods or wet places elsewhere. East central and south central Aroostook and northern Penobscot Counties and throughout, south of latitude 45.5, except central Somerset, northern Franklin, and northern Oxford Counties.

V. Opulus L. European Cranberry-bush.

See Rhod. 43:481-483. 1941; also Rehd. Man. ed. 2:842. 1940.

Native to Europe, northern Africa and northern Asia. Introduced and commonly cultivated as an ornamental and for its fruits. Occasionally escaping to and persisting in waste places near dwellings. Common in the urban centers. Throughout.

V. trilobum Marsh. Highbush Cranberry.

V. Opulus var. *americanum.* See Rhod. 43:481-483. 1941.

Frequent. Low woods and along the water courses generally; also thickets, roadsides, and other waste places as an escape. Often cultivated as an ornamental and for its fruits. Throughout, except southwestern Washington, northern and sontheastern Hancock, Lincoln, Sagadahoc, and eastern York Counties.

COMPOSITAE (COMPOSITE FAMILY)

Artemisia. Wormwood.

A. Abrotanum L. Southernwood.

Native to southern Europe. Introduced; cultivated and occasionally escaped. Locally established in West Poland, Androscoggin County and Biddeford Pool, York County.

A. Absinthium L. Wormwood.

Native to Europe. Introduced; locally thoroughly established and spreading along the highways, often distant from dwellings. East central Aroostook, southern Penobscot, Franklin, southern Oxford, southeastern Washington, southwestern Hancock, western Knox, western Androscoggin, and eastern Cumberland Counties.

A. pontica L. Roman Wormwood.

See Rehd. Man. ed. 2:883. 1940.

Native to Europe. Introduced; occasionally escaping from cultivation and becoming established near dwellings. Eastern Washington and along the coast in Cumberland and York Counties.

BIBLIOGRAPHY

The following list of references (with certain annotations) will be found useful in connection with the study of Maine woody plants. An asterisk (*) preceding the title denotes publications which contain the earliest known reports of Maine plants.

Bailey, L. H. 1938. Manual of cultivated plants. New York. Macmillan Co. 851 pp.
    Similar to Rehder's Manual but includes herbaceous plants of horticultural value.

——————. 1941, 1943, 1944. The genus Rubus in North America. Ithaca, N. Y. Bailey Hortorium, Cornell University. Gentes Herbarum 5: fasc. 1-7.

Baxter, James Phinney. 1890. *Sir Ferdinando Gorges and his province of Maine. Boston. John Wilson & Son. The publications of the Prince Society, Vol. 2.

Blake, Joseph. 1863. *A catalogue of the flowering plants of Maine. Proceedings of the Portland Society of Natural History 1:133-138.

Blake, S. F., and Alice C. Atwood. 1942. Geographical guide to floras of the world. Washington, D. C. U. S. Govt. Printing Office. U. S. Dept. Agric. Misc. Publ. No. 401. pp. 184-186.
    General and local references on Maine plants.

Briggs, F. P. 1892. *Flora of Mt. Katahdin. Plants collected at Mt. Katahdin, Maine. Bulletin of the Torrey Botanical Club 13:333-336.

Britton, Nathaniel L., and Addison Brown. 1913. Flora of the northern states and Canada. New York. Charles Scribner's Sons. 2nd ed. rev. 3 vols. 2052 pp.
    Contains keys and descriptions and is fully illustrated.

Champlain, Samuel de. 1907. *Voyages of Samuel de Champlain 1604-1618. Original narratives of early American history. New York. Charles Scribner's Sons.

Clapp, Roger. 1933. Woody plants for landscape planting in Maine. Orono, Me. University of Maine Studies, Second Series, No. 28. 91 pp.
    A bulletin listing 392 woody plants suitable for landscape planting and the zones in which they are hardy. Some useful information on plant material, hardiness, climate, and zoning is included.

Coburn, Louise Helen. 1920. Flora of Birch Island in Attean Pond. Rhodora 22:129-138.

Coburn, Louise Helen. Trees of Coburn Park, Skowhegan, Maine. Skowhegan, Me. Independent-Reporter Co. 73 pp.

Collins, J. Franklin, and Howard W. Preston. 1909. Key to New England trees, wild and commonly cultivated. Providence, R. I. Preston & Rounds Co.

Dame, L. L., and H. Brooks. 1904. Handbook of the trees of New England. Boston. Ginn & Co.

Dole, E. J. (Editor). 1937. The Flora of Vermont. Burlington, Vt. Free Press Printing Co. 3rd ed. rev. 353 pp.

 An annotated list of the ferns and seed plants of Vermont compiled by a committee of the Vermont Botanical Club.

Ewer, S. Judson. 1930. Notes on Katahdin plants. Rhodora 32:259-261.

Felter, Harvey Wickes. 1927. *The genesis of the American materia medica including a biographical sketch of "John Josselyn, Gent" and the medical and materia medica references in Josselyn's "New-England's Rarities Discovered," etc., and in his "Two Voyages to New-England." With critical notes and comments by H. W. Felter. Cincinnati, Ohio. Bulletin of the Lloyd library of botany, pharmacy and materia medica. Bulletin No. 26, Reproduction Series No. 8.

Fernald, M. L. 1892, 1895, and 1897. *The Portland catalogue of Maine plants. Proceedings of the Portland Society of Natural History 2: pt. 2, pp. 41-72; 2: pt. 3, pp. 73-96; 2: pt. 4, pp. 123-137.

―――――. 1902. The Relationships of some American and Old World Birches. Contributions of the Gray Herbarium, New Series, No. 22, reprinted without change of paging from American Journal of Science, Fourth Series, 14:167-194.

Fernald, M. L., and A. C. Kinsey. 1943. Edible wild plants of eastern North America. Cornwall-on-Hudson. N. Y. Idlewild Press. Illus. 452 pp.

 A special publication of the Gray Herbarium of Harvard University. A guide to all edible flowering plants and ferns, and some of the more important mushrooms, seaweeds, and lichens growing wild in the region east of the Great Plains and Hudson Bay and north of peninsular Florida.

Goodale, George Lincoln. 1861. *Botanical report. (In Sixth Annual Report of the Maine Board of Agriculture, pp. 125-129.)

―――――. 1861. *Botanical notes on the new lands. (In Sixth Annual Report of the Maine Board of Agriculture, pp. 361-372.)

―――――. 1862, 1863. *A catalogue of the flowering plants of Maine.

Proceedings of the Portland Society of Natural History 1:37-63; 127-133.

―――――. 1896. *New England plants seen by the earliest colonists. Publications of the Colonial Society of Massachusetts 3:180-194.

Harlow, William M. 1941. Twig key to the deciduous woody plants of eastern North America. Syracuse, N. Y. Published by the author. 4th rev. ed. 56 pp.

   A non-technical pocket-size key based upon winter twigs. Twigs and buds photographed and printed at enlargements of three diameters.

Harvey, F. L. 1893. *Catalogue of the North American phenogams and vascular cryptogams in the Blake herbarium. Orono, Me. I. Bulletin of the Maine State College Laboratory of Natural History 1: No. 2.

Harvey, F. L., and F. P. Briggs. 1893. *A contribution to the phenogams and vascular cryptogams of Maine. Orono, Me. II. Bulletin of the Maine State College Laboratory of Natural History 1: No. 2.

Harvey, LeRoy Harris. 1903. A study of the physiographic ecology of Mount Ktaadn, Maine. Orono, Me. University of Maine Studies, [First Series], No. 5:1-49.

   An early ecological study. (Out of print.)

―――――. 1909. The floristic composition of the vascular flora of Mount Katahdin, Maine. Eleventh report of the Michigan Academy of Science, pp. 32-47.

   Contains a table listing the 120 arctic alpine species of Mt. Katahdin and gives a discussion of the floristic affinities with arctic species of other parts of the world.

Hill, Albert Frederick. 1919. The vascular flora of the eastern Penobscot Bay region, Maine. Proceedings of the Portland Society of Natural History 3:199-304.

―――――. 1923. The vegetation of the Penobscot Bay region, Maine. Proceedings of the Portland Society of Natural History 3:305-438.

Hitchcock, C. H. 1862. *Geology of Maine. (In Seventh Annual Report of the Maine Board of Agriculture, pp. 223-430.)

Holmes, Ezekiel. 1861. *Agricultural adaptation of the wild lands explored. (In Sixth Annual Report of the Maine Board of Agriculture, pp. 356-360.)

Hyland, F. 1927. A report on the trees and native shrubs of Marsh Island, Orono and Old Town, Penobscot County, Maine. (Unpublished.)

Hyland, F. 1933. Our campus trees. Orono, Me. The Maine Alumnus 14, No. 7:109-110.
International rules of botanical nomenclature.... 1935. Revised by the International botanical congress of Cambridge, 1930; compiled by John Briquet. 3rd ed. Jena. G. Fischer.
Jackson, Charles T. 1837. *First report on the geology of the State of Maine. Augusta, Me. Smith & Robinson, printers. 128 pp.
Josselyn, John. 1672. *New-England's rarities discovered. (Reprinted with an introduction and notes by Edward Tuckerman. Transactions and collections of the American Antiquarian Society 4:105-238. 1860.)
————. 1675. *An account of two voyages to New-England. London. Printed for G. Widdowes. 2nd ed. 279 pp.

(This edition was reprinted in the Massachusetts Historical Society, Collections, Third Series, Vol. 3, 1833; and by William Veazie, Boston, Mass., in 1865.)

————. See also Felter, Harvey Wickes.
Josselyn Botanical Society of Maine. 1907-1939. Bulletin Nos. 1-7.

A bulletin covering proceedings of the Society.

Lamson-Scribner, F. 1875. *The ornamental and useful plants of Maine. Augusta, Me. Published by the author. 85 pp.

Contains popular descriptions and practical observations on the habits and properties of ornamental plants found native in the State.

————. 1891. Flora of Maine. A sketch of the flora of Orono, Maine. Botanical Gazette 16:228-234.

————. 1892. Mt. Katahdin and its flora. Botanical Gazette 17:46-54.
Lang, J. W. 1873. *List of trees and shrubs common to Waldo County, Maine. (In 18th Annual Report of the Maine Board of Agriculture, pp. 247-249.)
Lermond, Norman W. 1930. A list of trees and shrubs in Knox Arboretum, Thomaston, Maine.
Longfellow, A. W. *Geography, geology and botany of Cape Elizabeth, Maine. (In U. S. Coast Survey, Report of the Supt. pp. 31-32. 1853.)
Maine Forest Service. 1938. Forest Trees of Maine. Augusta, Me. 6th ed. (rev. by H. B. Peirson.) 86 pp.

A pocket-size pamphlet containing illustrations and descriptions of Maine trees. Seventy-six native species are included.

Maine Naturalist. 1921-1930. Vols. I-X.

A journal devoted to the fauna, flora, and geology of Maine.

Mathews, F. Schuyler. 1908. Field book of American wild flowers. New York. G. P. Putnam's Sons. 336 pp.

    A non-technical, pocket-size, illustrated handbook giving descriptions and general information; includes herbaceous plants and some shrubs.

―――――. 1915. Field book of American trees and shrubs. New York. G. P. Putnam's Sons.

    Similar to the above but includes only woody plants.

Moore, Barrington, and Norman Taylor. 1927. Vegetation of Mt. Desert Island, Maine, and its environment. Brooklyn Botanical Garden Memoirs, Vol. 3.

Muenscher, W. C. 1936. Key to woody plants. Ithaca, N. Y. Published by the author. 4th rev. ed. 105 pp.

    A non-technical pocket-size key based upon both summer and winter characters.

Norton, Arthur H. 1939. Bulletin of the Josselyn Botanical Society of Maine. No. 7:71-76.

    A bibliography of the Josselyn Botanical Society of Maine. An annotated list of 90 references on Maine plants, covering the period between 1895 and 1929.

Ogden, Eugene Cecil. 1934. The herbaceous flowering plants in the vicinity of Orono, Maine. Orono, Me. University of Maine Studies, Second Series, No. 34. 77 pp.

Pickering, John White. 1876. *Field notes in New England, York County, Maine. Field & Forest 2:43-44.

Portland Society of Natural History. 1862-1933. Portland, Maine. Proceedings Vol. 1, pt. 1—Vol. 4, pt. 2.

Rand, Edward L., and John H. Redfield. 1894. *Flora of Mt. Desert Island, Maine. Boston. John Wilson & Son. 286 pp.

    A preliminary catalog of plants growing on Mt. Desert and adjacent islands. Gives descriptions, ranges, characteristics, habits of growth, time of flowering and fruiting, and changes of foliage.

Rehder, Alfred. 1940. Manual of cultivated trees and shrubs hardy in North America exclusive of the subtropical and warmer temperate regions. New York. Macmillan Co. 2nd ed. 996 pp.

    A technical manual of the native and exotic trees and shrubs hardy in North America. Contains keys and descriptions, but no illustrations.

Rhodora. 1899 to date.

    The journal of the New England Botanical Club. A journal containing a variety of information on plants—expeditions, plant lists, monographs, illustrations and descriptions, keys, nomenclature, revisions, transfers, and occurrence and distribution. A necessary supplement to the manuals.

Robinson, B. L., and M. L. Fernald. 1908. Gray's new manual of botany. New York. American Book Co. 7th ed. illus. 926 pp.

    A technical handbook of the flowering plants and ferns of the central and northeastern United States and adjacent Canada. Contains keys, descriptions, and numerous illustrations.

Rousseau, Jacques. 1934. Essai de bibliographie botanique Canadienne. Montreal, Canada. Université de Montréal, Institut Botanique. I: Les travaux contenus dans Rhodora, Vols. 1-34, 1899-1932. (Reprinted from Naturaliste Canadien 60, 61:53-302.)

    A convenient index to Canadian plants listed in Rhodora Vol. 1 (1899) to Vol. 34 (1932).

St. John, Harold. 1929. Plants of the headwaters of the St. John River, Maine. Pullman, Washington. Research Studies of the State College of Washington 1:28-58.

    A list of plants collected or observed along the headwaters of the St. John River in northern Somerset and northwestern Aroostook Counties.

Sargent, Charles Sprague. 1922. Manual of the trees of North America (exclusive of Mexico). Boston. Houghton Mifflin Co. 2nd ed. 910 pp. 783 illustrations.

    A technical manual containing keys and descriptions of native trees only.

Shantz, H. L., et al. 1940. Approved changes in Sudworth's check list. Washington, D. C. U. S. Forest Service. 98 pp.

Sudworth, George B. 1927. Check list of the forest trees of the United States, their names and ranges. Washington, D. C. U. S. Dept. Agric. Misc. Cir. 92.

Taylor, William R. 1921. Flora of Mt. Desert, Maine. Additions to the flora of Mt. Desert. Rhodora 23:65-68.

Thoreau, Henry David. 1864. *Lists of plants native in the Maine woods in the years 1853 and 1857. (In his The Maine Woods. pp. 312-320.)

Wherry, Edgar T. 1928. Wild flowers of Mt. Desert Island, Maine. Published by the Garden Club of Mt. Desert.

    Contains several woody plants.

Wilkins, Austin H. 1932. The forests of Maine, their extent, character, ownership, and products. Augusta, Me. Bulletin No. 8. Maine Forest Service.

Williamson, William D. 1832. *History of the State of Maine 1:105-131. Hallowell, Maine. Glazier, Masters & Co.

    Contains an annotated list of trees, shrubs, vines, roots, and herbs.

Winship, George Parker. 1905. *Sailors narratives of voyages along the New England coast 1524-1624. Boston. Houghton Mifflin Co.

    A collection of voyages containing some of the earliest mention of the plants of New England.

INDEX TO THE FAMILIES, GENERA, AND COMMON NAMES INCLUDED IN THE CATALOG

Abies, 1
Acacia
  Rose-, 36
Acer, 40
ACERACEAE, 40
Aesculus, 41
Alder, 12
  American Green, 12
  Black-, 39
  Downy Green, 12
  Hoary, 12
  Smooth, 13
  Speckled, 12
  White-, 47
Alder-leaved Buckthorn, 41
Allegheny Serviceberry, 22
Allspice
  Wild, 19
Alnus, 12
Alpine Bearberry, 46
Alpine-azalea, 49
Alternate-leaved Dogwood, 45
Amelanchier, 21
American
  Aspen, 5
  Basswood, 43
  Beech, 15
  Cranberry, 51
  Elder, 56
  Elm, 17
  Fly Honeysuckle, 55
  Green Alder, 12
  Hazelnut, 14
  Hop-hornbeam, 14
  Hornbeam, 14
  Linden, 43
  Mountain-ash, 27
  Plane, 21
  Sycamore, 21
  Yew, 1
ANACARDIACEAE, 38
Andromeda, 46
Angelica-tree, 45
Appalachian Cherry, 26
Apple, 27, 28
  Crab, 28
  Thorn-, 23
AQUIFOLIACEAE, 39
Aralia, 45
ARALIACEAE, 45
Arbor-vitae, 4
  Eastern, 4

Arbutus
  Trailing, 47
Arceuthobium, 18
Arctostaphylos, 46
Aromatic Wintergreen, 47
Aronia, 27
Arrow-wood, 56, 57
  Maple-leaved, 56
Artemisia, 57
Ash, 52
  American Mountain-, 27
  Black, 52
  European Mountain-, 27
  Green, 53
  Mountain-, 27
  Northern Prickly-, 37
  Prickly-, 37
  Red, 53
  Wafer-, 37
  White, 52
Ash-leaved Maple, 40
Aspen, 4, 5
  American, 5
  Big-tooth, 5
  Large-tooth, 5
  Quaking, 5
  Small-tooth, 5
  Trembling, 5
Atlantic White-cedar, 3
Austrian Pine, 2
Azalea
  Alpine-, 49
  Clammy, 50

Balm-of-Gilead Poplar, 5
Balsam
  Fir, 1
  Poplar, 5
  Willow, 10
Baptisia, 36
Barberry, 19
  Common, 19
  Family, 19
  Japanese, 19
Basswood, 43
  American, 43
Bay Willow, 9
Bay-leaved Willow, 9
Bayberry, 11
Beach Plum, 26
Beaked Hazelnut, 14
Bean
  Indian-, 54

Bearberry, 46
  Alpine, 46
  Willow, 11
Bear Oak, 16
Bebb Willow, 7
Beech, 15
  American, 15
  Blue-, 14
  Copper, 15
  European, 15
  Family, 15
  Purple, 15
  Water-, 14
Benzoin, 19
BERBERIDACEAE, 19
Berberis, 19
Berry
  June-, 21
  Male-, 49
  Nanny-, 57
  Partridge-, 54
  Russet Buffalo-, 44
Betula, 13
BETULACEAE, 12
Bigleaf Shagbark
  Hickory, 11
Bignonia Family, 54
BIGNONIACEAE, 54
Big-tooth Aspen, 5
Bilberry, 50
  Bog, 52
  Dwarf, 51
  Mountain, 52
Birch, 13
  Black, 13
  Blue, 13
  Canoe, 13
  Cherry, 13
  Dwarf, 13
  Family, 12
  Gray, 14
  Low, 14
  Mountain Paper, 13
  Old-field, 14
  Paper, 13
  Poverty, 14
  Swamp, 14
  Sweet, 13
  White, 13
  Yellow, 13
Bird Cherry, 26
Bitter Nightshade, 54
Bittersweet
  Climbing, 40
  European, 54
  Shrubby, 40

65

Black
  Ash, 52
  Birch, 13
  Cherry, 26
  Crowberry, 37
  Currant, European, 20
  Currant, Swamp, 20
  Currant, Wild, 19
  Highbush Blueberry, 50
  Huckleberry, 48
  Locust, 37
  Oak, 17
  Raspberry, 34
  Spruce, 2
  Tupelo, 46
  Walnut, 12
  Willow, 9
Black-alder, 39
Black-berried Elder, 56
Black-fruited Choke-
    berry, 28
Black-gum, 46
Blackberry, 30
Blue
  Birch, 13
  Dogwood, 45
  Huckleberry, 48
  Spruce, 2
Blue-beech, 14
Blueberry, 50
  Black Highbush, 50
  Canada, 51
  Downy Swamp, 50
  Dryland, 52
  Early Sweet, 50
  Highbush, 51
  Late Low, 52
  Low Black-fruited
    Pale-leaved, 50
  Low Black-fruited
    Shiny-leaved, 50
  Low Sweet, 50
  Lowbush, 50
  Sour-top, 51
  Swamp, 51
  Velvet-leaf, 51
Bog
  Bilberry, 52
  Rosemary, 46
  Willow, 9
Bower
  Virgin's, 18
Box-elder, 40
Bramble, 30
Brier
  Common Green-, 4
  Sweet-, 29

Bristly
  Locust, 36
  Sarsaparilla, 45
Brittle Willow, 8
Broom, 36
  Crowberry, 37
  Scotch, 36
Buckeye, 41
  Ohio, 41
Buckthorn, 41
  Alder-leaved, 41
  Common, 41
  Family, 41
Buffalo-berry
  Russet, 44
Bullace Plum, 26
Bur Oak, 16
Burnett Rose, 30
Bush
  European Cranberry-,
    57
  Fever-, 19
  Hobble-, 56
  Shad-, 21
  Spice, 19
  Steeple, 35
Bush-honeysuckle, 55
Butternut, 12
Buttonbush, 54
  Common, 54
Buttonwood, 21

Calluna, 46
Canada
  Blueberry, 51
  Plum, 26
Canoe Birch, 13
Capillaire, 47
CAPRIFOLIACEAE, 55
Caragana, 36
Carpinus, 14
Carya, 11
Cashew Family, 38
Cassiope, 47
Castanea, 15
Catalpa, 54
Catalpa, 54
  Common, 54
  Northern, 54
  Southern, 54
Ceanothus, 41
Cedar
  Atlantic White-, 3
  Coast White-, 3
  Eastern Red-, 4
  Northern White-, 4
  Southern White-, 3
  White-, 3

CELASTRACEAE, 40
Celastrus, 40
Cephalanthus, 54
Chamaecyparis, 3
Chamaedaphne, 47
Checkerberry, 47
Cherry, 26
  Appalachian, 26
  Birch, 13
  Bird, 26
  Black, 26
  Common Choke, 27
  Fire, 26
  Pin, 26
  Rum, 26
  Sand, 26
  Sweet, 26
  Wild Red, 26
Chestnut, 15
  Horse-, 41
  Oak, 16
Chimaphila, 47
Chinese Elm, 17
Chinquapin Oak
  Dwarf, 16
Chiogenes, 47
Chokeberry, 27
  Black-fruited, 28
  Purple-fruited, 27
Choke Cherry
  Common, 27
Cigar-tree, 54
Cinnamon Rose, 29
Cinquefoil, 25
  Shrubby, 25
  Three-toothed, 25
CISTACEAE, 43
Cladrastis, 36
Clammy
  Azalea, 50
  Locust, 37
Clematis, 18
Clematis
  Purple, 18
Clethra, 47
Climbing
  Bittersweet, 40
  Honeysuckle, 55
  Nightshade, 54
Club
  Hercules', 45
Coast White-cedar, 3
Cockspur Thorn, 24
Coffee-tree
  Kentucky, 36
Common
  Barberry, 19
  Buckthorn, 41
  Buttonbush, 54

Common (Cont.)
  Catalpa, 54
  Choke Cherry, 27
  Eastern Dewberry, 32
  Elder, 56
  Green-brier, 4
  Hop-tree, 37
  Juniper, 3
  Lilac, 53
  Linden, 43
  Matrimony-vine, 53
  Osier, 11
COMPOSITAE, 57
Composite Family, 57
Comptonia, 11
Copper Beech, 15
Corema, 37
CORNACEAE, 45
Cornel
  Round-leaved, 46
  Silky, 45
Cornus, 45
Corylus, 14
Cottonwood, 4
  Eastern, 5
Crab
  Apple, 28
  Siberian, 27
Crack Willow, 8
Cranberry, 50
  American, 51
  Highbush, 57
  Large, 51
  Mountain, 52
  Rock, 52
  Small, 51
Cranberry-bush
  European, 57
Crataegus, 23
Creeper
  Virginia, 42
Creeping
  Juniper, 4
  Snowberry, 47
  Thyme, 53
Cricket-bat Willow, 6
Crowberry, 37
  Black, 37
  Broom, 37
  Family, 37
  Red, 37
Crowfoot Family, 18
Cucumber Magnolia, 18
Cucumber-tree, 18
CUPRESSACEAE, 3
Currant, 19
  European Black, 20
  Garden Red, 20
  Skunk, 20
  Swamp Black, 20
  Swamp Red, 20
  Wild Black, 19
Cypress Family, 3
Cytisus, 36

Daphne, 44
Decodon, 44
Devil's Walking-stick, 45
Dewberry, 30
  Common Eastern, 32
  Swamp, 32
Diapensia, 52
Diapensia Family, 52
DIAPENSIACEAE, 52
Diervilla, 55
Dirca, 44
Dockmackie, 56
Dogwood, 45
  Alternate-leaved, 45
  Blue, 45
  Family, 45
  Flowering, 45
  Gray, 46
  Pagoda, 45
  Panicled, 46
  Poison-, 39
  Red-osier, 46
  Round-leaved, 46
  Silky, 45
Douglas-fir, 3
Downy
  Green Alder, 12
  Serviceberry, 21
  Swamp Blueberry, 50
Dryland Blueberry, 52
Dwarf
  Bilberry, 51
  Birch, 13
  Chinquapin Oak, 16
  Gray Willow, 10
  Huckleberry, 48
  Juniper, 4
  Mistletoe, 18
  Sumac, 38
  Willow, 8
Dyer's Greenweed, 36

Early Sweet Blueberry, 50
Eastern
  Arbor-vitae, 4
  Cottonwood, 5
  Hemlock, 3
  Hop-hornbeam, 14
  Larch, 1
  Red Oak, 16
  Red-cedar, 4
  White Pine, 3
Eglantine, 29
ELAEAGNACEAE, 44
Elaeagnus, 44
Elder, 56
  American, 56
  Black-berried, 56
  Box-, 40
  Common, 56
  Red-berried, 56
Elm, 17
  American, 17
  Chinese, 17
  Family, 17
  Red, 17
  Scotch, 17
  Siberian, 17
  Slippery, 17
  White, 17
  Wych, 17
EMPETRACEAE, 37
Empetrum, 37
English
  Hawthorn, 25
  Oak, 17
Epigaea, 47
ERICACEAE, 46
European
  Beech, 15
  Bittersweet, 54
  Black Currant, 20
  Cranberry-bush, 57
  Fly Honeysuckle, 56
  Gooseberry, 20
  Mountain-ash, 27
  Small-leaved Linden, 43

FAGACEAE, 15
Fagus, 15
False Indigo, 36
False-spiraea
  Ural, 35
Fern
  Sweet-, 11
Fever-bush, 19
Fir, 1
  Balsam, 1
  Douglas-, 3
  White, 1
Fire Cherry, 26
Flowering
  Dogwood, 45
  Raspberry, Purple, 34
Fly Honeysuckle
  American, 55
  European, 56
  Mountain, 55
  Swamp, 55
Fox Grape
  Northern, 42

Fraxinus, 52
French Rose, 29
Frost Grape, 42
Frostweed, 43
  Hoary, 43

Gale
  Sweet, 11
Garden Red Currant, 20
Gaultheria, 47
Gaylussacia, 48
Genista, 36
Ginkgo, 1
Ginkgo, 1
  Family, 1
GINKGOACEAE, 1
Ginseng Family, 45
Glaucous Willow, 7
Gleditsia, 36
Golden Willow, 6
Gooseberry, 19
  European, 20
  Prickly, 19
  Smooth, 20
Goose-foot Maple, 40
Grape, 42
  Frost, 42
  New England, 42
  Northern Fox, 42
  Pilgrim, 42
  River-bank, 42
Gray
  Birch, 14
  Dogwood, 46
Great-laurel, 49
Green Ash, 53
Green-brier, 4
  Common, 4
Greenweed
  Dyer's, 36
Ground Juniper, 4
Ground-hemlock, 1
Ground-pink, 53
Gum
  Black-, 46
  Sour-, 46
Gymnocladus, 36

HAMAMELIDACEAE, 20
Hamamelis, 20
Hard Maple, 41
Hardhack, 35
Hawthorn, 23
  English, 25
Hazelnut, 14
  American, 14
  Beaked, 14

Heath
  Family, 46
  Mountain, 49
Heath-like Hudsonia, 43
Heather, 46
Helianthemum, 43
Hemlock, 3
  Eastern, 3
  Ground-, 1
Hercules' Club, 45
Hickory, 11
  Bigleaf Shagbark, 11
  Shagbark, 12
  Shellbark, 11
Hicoria, 11
Highbush
  Blueberry, 51
  Blueberry, Black, 50
  Cranberry, 57
Hoary
  Alder, 12
  Frostweed, 43
  Willow, 7
Hobble
  Witch, 56
Hobble-bush, 56
Holly, 39
  Family, 39
  Mountain, 39
Honey-locust, 36
Honeysuckle, 55
  American Fly, 55
  Bush-, 55
  Climbing, 55
  European Fly, 56
  Family, 55
  Morrow, 55
  Mountain Fly, 55
  Swamp Fly, 55
  Tatarian, 55
  Trumpet, 55
Hop-hornbeam, 14
  American, 14
  Eastern, 14
Hop-tree, 37
  Common, 37
Hornbeam, 14
  American, 14
Horse Plum, 26
Horse-chestnut, 41
Huckleberry, 48
  Black, 48
  Blue, 48
  Dwarf, 48
  White, 48
Hudsonia, 43
Hudsonia
  Heath-like, 43

  Woolly, 43

Ilex, 39
Indian-bean, 54
Indigo
  False, 36
  Wild, 36
Inkberry, 39
Ironwood, 14
Ivy
  Poison, 38

Jack Pine, 2
Japanese Barberry, 19
JUGLANDACEAE, 11
Juglans, 12
June-berry, 21
Juniper, 1, 3
  Common, 3
  Creeping, 4
  Dwarf, 4
  Ground, 4
  Mountain, 4
  Prostrate, 4
Juniperus, 3

Kalmia, 48
Kentucky Coffee-tree, 36
King Nut, 11

LABIATAE, 53
Labrador Tea, 49
Lamb-kill, 48
Lapland
  Rhododendron, 49
  Rose-bay, 49
Larch, 1
  Eastern, 1
Large Cranberry, 51
Large-tooth Aspen, 5
Larix, 1
Late Low Blueberry, 52
LAURACEAE, 19
Laurel, 48
  Family, 19
  Great-, 49
  Mountain-, 48
  Pale-, 49
  Sheep-, 48
Leather-leaf, 47
Leatherwood, 44
Ledum, 49
LEGUMINOSAE, 36
Lepargyraea, 44
Leverwood, 14
Lilac, 53
  Common, 53

LILIACEAE, 4
Lily Family, 4
Linden, 43
　American, 43
　Common, 43
　Family, 43
　Small-leaved European, 43
Lindera, 19
Linnaea, 55
Liriodendron, 18
Locust, 36
　Black, 37
　Bristly, 36
　Clammy, 37
　Honey-, 36
Loiseleuria, 49
Lombardy Poplar, 5
Lonicera, 55
Loosestrife
　Family, 44
　Swamp, 44
LORANTHACEAE, 18
Low
　Birch, 14
　Black-fruited Pale-leaved Blueberry, 50
　Black-fruited Shiny-leaved Blueberry, 50
　Sweet Blueberry, 50
　Lowbush Blueberry, 50
Lowell Oak, 16
Lycium, 53
Lyonia, 49
LYTHRACEAE, 44

Madder Family, 54
Magnolia, 18
Magnolia
　Cucumber, 18
　Family, 18
MAGNOLIACEAE, 18
Maidenhair-tree, 1
Male-berry, 49
Malus, 27
Maple, 40
　Ash-leaved, 40
　Family, 40
　Goose-foot, 40
　Hard, 41
　Mountain, 41
　Norway, 40
　Red, 40
　Rock, 41
　Schwedler, 40
　Silver, 40
　Striped, 40
　Sugar, 41

Sycamore, 40
　White, 40
　Wier, 41
Maple-leaved Arrow-wood, 56
Matrimony-vine, 53
　Common, 53
Mayflower, 47
Mazzard, 26
Meadowsweet, 35
Mezereum, 44
　Family, 44
Mint Family, 53
Mistletoe
　Dwarf, 18
　Family, 18
Mitchella, 54
Mooseberry, 57
Moosewood, 40, 56
MORACEAE, 18
Morrow Honeysuckle, 55
Morus, 18
Moss-pink, 53
Mountain
　Bilberry, 52
　Cranberry, 52
　Fly Honeysuckle, 55
　Heath, 49
　Holly, 39
　Juniper, 4
　Maple, 41
　Paper Birch, 13
　Pine, 2
Mountain-ash, 27
　American, 27
　European, 27
Mountain-laurel, 48
Moxie-plum, 47
Mugo Pine, 2
Mulberry, 18
　Family, 18
　White, 18
Myrica, 11
Myrica, 11
MYRICACEAE, 11

Nanny-berry, 57
Nemopanthus, 39
New England Grape, 42
New Jersey Tea, 41
Nightshade, 54
　Bitter, 54
　Climbing, 54
　Family, 53
Nine-bark, 25
Northern
　Catalpa, 54
　Fox Grape, 42

Prickly-ash, 37
Red Oak, 15
White Pine, 3
White-cedar, 4
Norway
　Maple, 40
　Pine, 2
　Spruce, 1
Nut
　King, 11
Nyssa, 46

Oak, 15
　Bear, 16
　Black, 17
　Bur, 16
　Chestnut, 16
　Dwarf Chinquapin, 16
　Eastern Red, 16
　English, 17
　Lowell, 16
　Northern Red, 15
　Pin, 16
　Poison-, 38
　Porter, 16
　Red, 15, 16
　Scarlet, 16
　Swamp White, 15
　White, 15
Ohio Buckeye, 41
Old-field Birch, 14
OLEACEAE, 52
Oleaster, 44
　Family, 44
Olive
　Family, 52
　Russian, 44
Osier
　Common, 11
　Purple, 10
Ostrya, 14

Pagoda Dogwood, 45
Pale-laurel, 49
Panicled Dogwood, 46
Paper Birch, 13
　Mountain, 13
Parthenocissus, 42
Partridge-berry, 54
Pea-tree, 36
Pear, 27
Pepperbush
　Sweet, 47
Pepperidge, 46
Phlox, 53
Phyllodoce, 49
Physocarpus, 25
Picea, 1

Pignut
 Sweet, 11
Pilgrim Grape, 42
Pimbina, 57
Pin
 Cherry, 26
 Oak, 16
PINACEAE, 1
Pine, 2
 Austrian, 2
 Eastern White, 3
 Family, 1
 Jack, 2
 Mountain, 2
 Mugo, 2
 Northern White, 3
 Norway, 2
 Pitch, 3
 Prince's-, 47
 Pumpkin, 3
 Red, 2
 Scotch, 3
 Soft, 3
 Umbrella-, 3
Pink
 Ground-, 53
 Moss-, 53
Pinus, 2
Pipsissewa, 47
Pitch Pine, 3
Plane
 American, 21
 Tree Family, 21
PLATANACEAE, 21
Platanus, 21
Plum, 26
 Beach, 26
 Bullace, 26
 Canada, 26
 Horse, 26
 Moxie-, 47
 Red, 26
 Sugar-, 21
 Wild, 26
Poison
 Ivy, 38
 Sumac, 39
Poison-dogwood, 39
Poison-oak, 38
POLEMONIACEAE, 53
Polemonium Family, 53
Poplar, 4, 5
 Balm-of-Gilead, 5
 Balsam, 5
 Lombardy, 5
 White, 4
 Yellow-, 18
Popple, 5

Populus, 4
Porter Oak, 16
Potentilla, 25
Poverty Birch, 14
Prairie Willow, 8
Prickly
 Gooseberry, 19
 Sarsaparilla, 45
Prickly-ash, 37
 Northern, 37
Prince's-pine, 47
Prostrate Juniper, 4
Prunus, 26
Psedera, 42
Pseudotsuga, 3
Ptelea, 37
Pulse Family, 36
Pumpkin Pine, 3
Purple
 Beech, 15
 Clematis, 18
 Flowering Raspberry, 34
 Osier, 10
 Willow, 10
Purple-fruited Choke-
 berry, 27
Pussy Willow, 7
Pyrus, 27

Quaking Aspen, 5
Quercus, 15

Raisin
 Wild, 56, 57
RANUNCULACEAE, 18
Raspberry, 30
 Black, 34
 Purple Flowering, 34
 Red, 32
 Wild Red, 32
Red
 Ash, 53
 Cherry, Wild, 26
 Crowberry, 37
 Currant, Garden, 20
 Currant, Swamp, 20
 Elm, 17
 Maple, 40
 Oak, 15, 16
 Oak, Eastern, 16
 Oak, Northern, 15
 Pine, 2
 Plum, 26
 Raspberry, 32
 Raspberry, Wild, 32
 Spruce, 2

Red-berried Elder, 56
Red-cedar
 Eastern, 4
Red-osier Dogwood, 46
Red-root, 41
Redwood Family, 3
RHAMNACEAE, 41
Rhamnus, 41
Rhododendron, 49
Rhododendron
 Lapland, 49
 Rose-bay, 49
Rhodora, 49
Rhus, 38
Ribes, 19
River-bank Grape, 42
Robinia, 36
Rock
 Cranberry, 52
 Maple, 41
Rockrose, 43
 Family, 43
Roman Wormwood, 57
Root
 Red-, 41
Rosa, 28
ROSACEAE, 21
Rose, 28
 Burnett, 30
 Cinnamon, 29
 Family, 21
 French, 29
 Rugosa, 30
 Scotch, 30
Rose-acacia, 36
Rose-bay
 Lapland, 49
 Rhododendron, 49
Rosemary
 Bog, 46
Round-leaved
 Cornel, 46
 Dogwood, 46
Roundwood, 27
Rowan Tree, 27
RUBIACEAE, 54
Rubus, 30
Rue Family, 37
Rugosa Rose, 30
Rum Cherry, 26
Russet Buffalo-berry, 44
Russian Olive, 44
RUTACEAE, 37

Sage Willow, 7
SALICACEAE, 4
Salix, 6
Sambucus, 56

Sand Cherry, 26
Sandbar Willow, 8
SAPINDACEAE, 41
Sarsaparilla
  Bristly, 45
  Prickly, 45
  Wild, 45
Sassafras, 19
Sassafras, 19
SAXIFRAGACEAE, 19
Saxifrage Family, 19
Scarlet Oak, 16
Schwedler Maple, 40
Sciadopitys, 3
Scotch
  Broom, 36
  Elm, 17
  Pine, 3
  Rose, 30
Serviceberry, 21
  Allegheny, 22
  Downy, 21
Shad-bush, 21
Shagbark Hickory, 12
Sheep-laurel, 48
Shellbark Hickory, 11
Shepherdia, 44
Shining
  Sumac, 38
  Willow, 8
Shrubby
  Bittersweet, 40
  Cinquefoil, 25
  Trefoil, 37
Siberian
  Crab, 27
  Elm, 17
Silky
  Cornel, 45
  Dogwood, 45
  Willow, 10
Silver
  Maple, 40
  Pussy Willow, 7
Skunk Currant, 20
Slippery Elm, 17
Small Cranberry, 51
Small-leaved European
    Linden, 43
Small-tooth Aspen, 5
Smilax, 4
Smooth
  Alder, 13
  Gooseberry, 20
  Sumac, 38
  Winterberry, 39
Snowberry, 56
  Creeping, 47

Soapberry Family, 41
Soft Pine, 3
SOLANACEAE, 53
Solanum, 54
Sorbaria, 35
Sorbaronia, 27
Sorbus, 27
Sour-gum, 46
Sour-top Blueberry, 51
Southern
  Catalpa, 54
  White-cedar, 3
Southernwood, 57
Speckled Alder, 12
Spice-bush, 19
Spiraea, 35
Spiraea, 35
  Ural False-, 35
Spruce, 1
  Black, 2
  Blue, 2
  Norway, 1
  Red, 2
  White, 2
Squashberry, 57
Staff-tree, 40
  Family, 40
Staghorn Sumac, 39
Steeple Bush, 35
Striped Maple, 40
Sugar Maple, 41
Sugar-plum, 21
Sumac, 38
  Dwarf, 38
  Poison, 39
  Shining, 38
  Smooth, 38
  Staghorn, 39
Swamp
  Birch, 14
  Black Currant, 20
  Blueberry, 51
  Dewberry, 32
  Fly Honeysuckle, 55
  Loosestrife, 44
  Red Currant, 20
  White Oak, 15
Sweet
  Birch, 13
  Cherry, 26
  Gale, 11
  Gale Family, 11
  Pepperbush, 47
  Pignut, 11
  Viburnum, 57
Sweet-brier, 29
Sweet-fern, 11
Sycamore, 21

American, 21
  Maple, 40
Symphoricarpos, 56
Syringa, 53

Tamarack, 1
Tatarian Honeysuckle, 55
TAXACEAE, 1
TAXODIACEAE, 3
Taxus, 1
Tea
  Labrador, 49
  New Jersey, 41
Teaberry, 47
Thorn, 23
Thorn-apple, 23
  Cockspur, 24
Three-toothed Cinquefoil,
    25
Thuja, 4
Thyme, 53
  Creeping, 53
THYMELAEACEAE, 44
Thymus, 53
Tilia, 43
TILIACEAE, 43
Toothache-tree, 37
Trailing Arbutus, 47
Tree
  Angelica-, 45
  Cigar-, 54
  Common Hop-, 37
  Cucumber-, 18
  Hop-, 37
  Kentucky Coffee-, 36
  Pea-, 36
  Rowan-, 27
  Staff-, 40
  Toothache-, 37
  Tulip-, 18
Trefoil
  Shrubby, 37
Trembling Aspen, 5
Trumpet Honeysuckle, 55
Tsuga, 3
Tulip-tree, 18
Tupelo, 46
  Black, 46
Twin-flower, 55

ULMACEAE, 17
Ulmus, 17
Umbrella-pine, 3
Ural False-spiraea, 35

Vaccinium, 50
Velvet-leaf Blueberry, 51
Viburnum, 56

Viburnum
　Sweet, 57
Vine
　Common Matrimony-, 53
　Family, 42
　Matrimony-, 53
Virginia Creeper, 42
Virgin's Bower, 18
VITACEAE, 42
Vitis, 42

Wafer-ash, 37
Walking-stick
　Devil's, 45
Walnut, 12
　Black, 12
　Family, 11
　White, 12
Water-beech, 14
Water-willow, 44
Waxwork, 40
Weeping Willow, 6
Whin, 36
White
　Ash, 52
　Birch, 13
　Elm, 17
　Fir, 1
　Huckleberry, 48
　Maple, 40
　Mulberry, 18
　Oak, 15
　Poplar, 4
　Spruce, 2
　Walnut, 12
　Willow, 6

White-alder, 47
White-cedar, 3
　Atlantic, 3
　Coast, 3
　Northern, 4
　Southern, 3
White-wood, 18
Wicopy, 44
Wier Maple, 41
Wild
　Allspice, 19
　Black Currant, 19
　Indigo, 36
　Plum, 26
　Raisin, 56, 57
　Red Cherry, 26
　Red Raspberry, 32
　Sarsaparilla, 45
Willow, 6
　Balsam, 10
　Bay, 9
　Bay-leaved, 9
　Bearberry, 11
　Bebb, 7
　Black, 9
　Bog, 9
　Brittle, 8
　Crack, 8
　Cricket-bat, 6
　Dwarf, 8
　Dwarf Gray, 10
　Family, 4
　Glaucous, 7
　Golden, 6
　Hoary, 7
　Prairie, 8
　Purple, 10
　Pussy, 7

Sage, 7
Sandbar, 8
Shining, 8
Silky, 10
Silver Pussy, 7
Water-, 44
Weeping, 6
White, 6
Winterberry, 39
　Smooth, 39
Wintergreen
　Aromatic, 47
Witch Hobble, 56
Witch-hazel, 20
　Family, 20
Withe-rod, 56
Woad-waxen, 36
Wood
　White-, 18
　Yellow-, 36
Woodbine, 42
Woolly Hudsonia, 43
Wormwood, 57
　Roman, 57
Wych Elm, 17

Xanthoxylum, 37

Yellow Birch, 13
Yellow-poplar, 18
Yellow-wood, 36
Yew, 1
　American, 1
　Family, 1

Zanthoxylum, 37